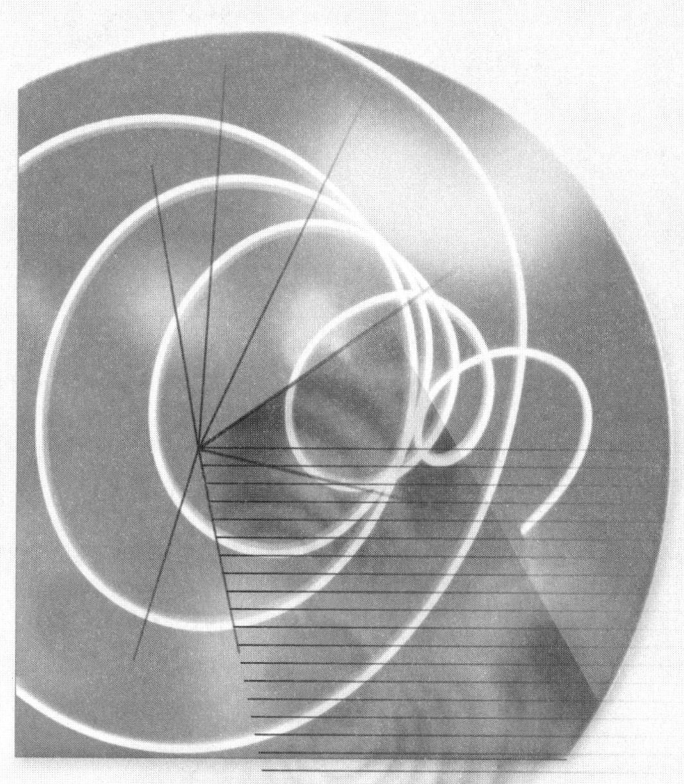

AutoCAD Database Connectivity

AutoCAD Database Connectivity

Autodesk Press

Thomson Learning™

Africa • Australia • Canada • Denmark • Japan • Mexico • New Zealand Phillipines • Puerto Rico • Singapore • Spain • United Kingdom • United States

NOTICE TO THE READER

Trademarks

Autodesk Press Staff
Executive Director: Alar Elken
Executive Editor: Sandy Clark
Development: Allyson Powell
Executive Marketing Manager: Maura Theriault
Executive Production Manager: Mary Ellen Black
Production Coordinator: Jennifer Gaines
Art and Design Coordinator: Mary Beth Vought
Marketing Coordinator: Paula Collins
Technology Project Manager: Tom Smith

Cover illustration by Brucie Rosch

Printed in Canada
2 3 4 5 6 7 8 9 10 XXX 04 03 02 01 00

For more information, contact
Autodesk Press
3 Columbia Circle, Box 15-015
Albany, New York USA 12212-15015;
or find us on the World Wide Web at http://www.autodeskpress.com

Library of Congress Cataloging-in-Publication Data

McFarlane, Scott.
AutoCAD database connectivity / Scott McFarlane.
p. cm.
ISBN 0-7668-1640-0 (pbk.)
1. Computer graphics. 2. AutoCAD. I. Title.
T385.M3816 1999
620'.0042'02855369—dc21 99-026737

CONTENTS

FOREWORD

A WORD FROM THOMSON LEARNING™ xiii
A WORD FROM AUTODESK, INC. xiii
THE INITIAL IDEA xiv
TOPICS FOR THE PROGRAMMER SERIES xv
WHO READS PROGRAMMING BOOKS? xv
THIRST, THEME, AND VARIATION xvi

PREFACE

WHY CONNECT AUTOCAD™ TO DATABASES? xvii
PURPOSE OF THIS BOOK xviii
WHO SHOULD READ THIS BOOK xviii
WHAT YOU NEED FOR THIS BOOK xviii
HOW THIS BOOK IS ORGANIZED xix
WHAT'S NOT COVERED AND WHY xix
HOW TO USE THIS BOOK xx
CONVENTIONS USED INTHIS BOOK xx
- Tutorials xx
- Code Syntax xx
- Example Code xx

FEATURES xxi
USING THE CD xxi
WE WANT TO HEAR FROM YOU! xxii
ABOUT THE AUTHOR xxii
A WORD ABOUT RELATED TECHNOLOGY TRENDS xxii
ACKNOWLEDGMENTS xxiii

CHAPTER 1—AN INTRODUCTION TO DBCONNECT

OBJECTIVES 1
INTRODUCTION 1
- A Brief History 1
- Example Applications 2

GETTING STARTED 3
- dbConnect and ASE 3
- Basic dbConnect Features 3

User Interface Components ... 3
Invoking dbConnect Commands ... 3
The dbConnect Manager ... 4
ESTABLISHING A DATABASE CONNECTION ... 6
Data Links and Data Sources ... 6
TUTORIAL 1.1 – CONFIGURING A DATA SOURCE ... 6
Connecting to the Database ... 9
VIEWING AND EDITING TABLE DATA ... 9
LINKING OBJECTS TO THE DATABASE ... 11
Creating Link Templates ... 12
TUTORIAL 1.2 – CREATING LINKS TO GRAPHICAL OBJECTS ... 12
CREATING LABELS ... 14
Creating Label Templates ... 15
TUTORIAL 1.3 – CREATING FREESTANDING LABELS ... 15
THE QUERY EDITOR ... 17
Quick Query ... 18
TUTORIAL 1.4 – CREATING A SIMPLE QUERY ... 18
Range Query ... 20
TUTORIAL 1.5 – CREATING A RANGE QUERY ... 20
SUMMARY ... 21
REVIEW QUESTIONS ... 22

CHAPTER 2—ADVANCED DBCONNECT FEATURES

OBJECTIVES ... 23
INTRODUCTION ... 23
ADVANCED DATA VIEW FEATURES ... 23
The Data View Interface ... 24
Data Manipulation ... 27
TUTORIAL 2.1 – MODIFYING DATA IN THE DATA VIEW WINDOW ... 29
Changing the Appearance of the Data View Window ... 30
Viewing Linked Records and Linked Objects ... 32
Controlling Data View Options ... 33
TUTORIAL 2.2 – VIEWING LINKED RECORDS AND LINKED OBJECTS ... 35
Printing the Data View ... 35
Using the Clipboard ... 36
CREATING COMPLEX QUERIES ... 36
The Query Builder ... 36
TUTORIAL 2.3 – USING THE QUERY BUILDER ... 37
SQL Query ... 43
TUTORIAL 2.4 – USING THE SQL QUERY TAB ... 43
Query Processing Options ... 45
Sharing Queries with Other Drawings ... 45
SELECTING OBJECTS USING LINK SELECT ... 46
TUTORIAL 2.5 – USING LINK SELECT ... 48

EXPORTING LINKS....50
TUTORIAL 2.6 – EXPORTING LINKS TO A DATABASE TABLE....51
MANAGING LINK TEMPLATES....53
Modifying a Link Template....53
Deleting a Link Template....56
Sharing Link Templates with Other Drawings....56
CHECKING LINK INTEGRITY....57
TUTORIAL 2.7 – SYNCHRONIZING LINKS....58
USING OBJECT SHORTCUT MENUS....59
CONVERTING OLD LINKS....61
TUTORIAL 2.8 – CONVERTING OLD LINKS....61
LINKING TO OTHER DATABASE SYSTEMS....63
Using Microsoft Excel....63
TUTORIAL 2.9 – DEFINING A NAMED RANGE OF CELLS IN EXCEL....64
TUTORIAL 2.10 – CREATING THE ODBC DATA SOURCE....65
TUTORIAL 2.11 – MODIFYING THE PARCELS DATA SOURCE IN AUTOCAD....66
Microsoft SQL Server and Oracle....68
GLOBAL DBCONNECT OPTIONS....68
SUMMARY....69
REVIEW QUESTIONS....69

CHAPTER 3—DATABASE DESIGN

OBJECTIVES....71
INTRODUCTION....71
Why Is It So Important?....72
How Databases Are Used....73
Goals of Database Design....73
The Relational Data Model....74
ENTITY RELATIONSHIP (ER) DIAGRAMS....74
Components....74
Weak Entities....75
THE DESIGN PROCESS....75
Establishing User Requirements....76
Identifying and Defining Entities....77
Identifying Relationships....78
Identifying Attributes....79
Identifying Primary and Foreign Keys....80
Normalization....83
DESIGNING THE CONFERENCE DATABASE....88
Establishing User Requirements....88
Identifying and Defining Entities....89
Identifying Relationships....89
Identifying Attributes....90

Identifying Primary and Foreign Keys 92
Normalization....94
Referential Integrity 96
Other Data Integrity Issues....98
Validating the Design....98
SUMMARY 99
REVIEW QUESTIONS 99

CHAPTER 4—USING SQL

OBJECTIVES....101
INTRODUCTION....101
The Example Conference Database....102
Executing the SQL Examples....103
TUTORIAL 4.1 – CONFIGURING A DATA SOURCE FOR THE CLASS DATABASE....103
THE BASIC RULES OF SQL 103
The SQL Hierarchy 103
SQL Data Types 105
Types of SQL Commands 106
USING SQL TO CREATE QUERIES 107
The SELECT statement 107
The WHERE Clause....108
Conditional Expressions 108
Sorting Output with ORDER BY 112
Querying from Multiple Tables....112
Aggregate Functions 114
ADVANCED QUERY CONCEPTS....118
Using Views....118
Querying a Table Against Itself....121
Embedding a Query Inside Another Query....123
USING SQL TO MODIFY THE DATABASE 124
The INSERT Command....124
The UPDATE Command 126
The DELETE Command 126
USING SQL QUERIES FOR DATA VALIDATION....127
PUTTING SQL TO WORK IN AUTOCAD....129
Deciding Where to Make the Link....129
Linking a Drawing to the Class Database....130
TUTORIAL 4.2 – LINKING A DRAWING TO THE CLASS DATABASE 131
Graphically Showing Query Results....131
Handling One-to-Many Relationships....133
Using Labels....133
TUTORIAL 4.3 – CREATING THE ROOM_VIEW LINK TEMPLATE....136
TUTORIAL 4.4 – CREATING THE FIRST ROOM_VIEW LABEL TEMPLATE ..136

TUTORIAL 4.5 – CREATING THE SECOND ROOM_VIEW LABEL TEMPLATE...... 136
TUTORIAL 4.6 – CREATING THE LABELS 137
TUTORIAL 4.7 – CREATING THE INTERFACE 138
SUMMARY 139
REVIEW QUESTIONS 139
EXERCISES 139

CHAPTER 5—DESIGNING AUTOCAD/DATABASE APPLICATIONS

OBJECTIVES 141
INTRODUCTION 141
Example Asset Management Application 142
WHY DO YOU NEED A CUSTOM APPLICATION? 142
Providing a Graphical User Interface 142
Dealing with Dual Environments 143
USING OFF-THE-SHELF APPLICATIONS 143
INTERNAL VERSUS EXTERNAL STORAGE OF DATA 143
Internal Storage 144
External Storage 147
How Do You Decide? 149
TYPES OF CAD/DATABASE APPLICATIONS 150
LINKING SCHEMES 152
Many-to-One or One-to-Many 152
Many-to-Many 152
One-to-One 154
GUIDELINES FOR A SUCCESSFUL APPLICATION 155
DESIGNING AN ASSET MANAGEMENT APPLICATION 156
The Problem Statement 157
Existing Conditions and Needs Assessment 157
Database Design 159
Identifying Custom Applications 166
Developing the Applications 168
SUMMARY 168
REVIEW QUESTIONS 168
EXERCISES 169

CHAPTER 6—ACTIVEX DATA OBJECTS (ADO)

OBJECTIVES 171
INTRODUCTION 171
The Component Object Model (COM) 171
Files Needed for this Chapter 172
THE ADO OBJECT MODEL 172
OVERVIEW OF ADO 174
A Typical Database Transaction 174

USING ADO WITH AUTOCAD VBA 177
Referencing the ADO Library 177
TUTORIAL 6.1 – REFERENCING THE ADO LIBRARY IN VBA 177
Iterating Through A Recordset 178
Modifying Data in the Recordset 179
Adding New Rows to a Table 180
Putting ADO to Work in AutoCAD 180
Exporting Drawing Information to a Database 180
TUTORIAL 6.2 – EXPORTING DRAWING INFORMATION TO A DATABASE 180
Updating the Drawing from the Database 186
TUTORIAL 6.3 – UPDATING THE DRAWING FROM THE DATABASE 186
Searching for Specific Rows in the Recordset 188
TUTORIAL 6.4 – UPDATING THE DATABASE FROM THE DRAWING 189
Getting the List of Available Data Sources 191
Getting the List of Available Tables in a Data Source 192
Executing SQL Commands 193
USING ADO WITH VISUAL LISP 193
Accessing COM Libraries 193
Importing the ADO Library 194
Creating an Instance of an ADO Object 197
Error Trapping 197
Retrieving a Recordset 199
Storing Drawing Data in a Database 201
SUMMARY 204
REVIEW QUESTIONS 205
EXERCISE 205

CHAPTER 7—CONNECTIVITY AUTOMATION OBJECTS (CAO)

OBJECTIVES 207
INTRODUCTION 207
Files Needed for this Chapter 208
THE CAO OBJECT MODEL 208
USING THE CAO LIBRARY IN VBA 209
OVERVIEW OF THE CAO LIBRARY 209
The DbConnect Object 209
The LinkTemplates Collection 209
The Links Collection 214
Errors 215
PUTTING THE CAO LIBRARY TO WORK 216
Getting Link Information 216
Creating a New Link on an Object 219
Modifying an Existing Link 221
Deleting a Link from an Object 222

Reloading Labels 224
Capturing Errors 225
Selecting Linked Objects 226
SUMMARY 228
REVIEW QUESTIONS 229
EXERCISES 229

CHAPTER 8—PUTTING IT ALL TOGETHER

OBJECTIVES 231
INTRODUCTION 231
THE ASSET MANAGEMENT APPLICATION REVISITED 231
Set Up the Application Files 232
TUTORIAL 8.1 – EXAMINING THE DRAWING 232
TUTORIAL 8.2 – CONFIGURING THE DATA SOURCE 234
TUTORIAL 8.3 – CREATING THE LINK TEMPLATE 235
UNDERSTANDING THE RELATIONSHIP BETWEEN CAO AND ADO 238
A Typical Scenario 238
DEVELOPING CAO UTILITY FUNCTIONS 241
DEVELOPING THE APPLICATIONS 243
Data Creation Applications 243
Data Maintenance Applications 250
Data Integrity Validation Applications 252
Query and Annotation Applications 253
SUMMARY 258

GLOSSARY 259

APPENDIX A—DBCONNECT COMMAND REFERENCE

DBCONNECT COMMANDS 265
DATA VIEW COMMANDS 269

APPENDIX B—CHANGES FROM ASE

INTRODUCTION 271
TERMINOLOGY CHANGES 271
COMMAND CHANGES 272
PROGRAMMING INTERFACE CHANGES 272
ASI AutoLISP Interface 272
ASI Link AutoLISP Interface 275

INDEX 277

FOREWORD

A WORD FROM THOMSON LEARNING™

Autodesk Press was formed in 1995 as a global strategic alliance between Thomson Learning and Autodesk, Inc. We are pleased to bring you the premier publishing list of Autodesk student software and learning and training materials to support the Autodesk family of products. AutoCAD™ is such a powerful product that everyone who uses it would benefit from a mentor to help them unlock its full potential. This is the premise upon which the Programmer Series was conceived. The titles in this series cover the most advanced topics that will help you maximize AutoCAD. Our Programmer Series titles also bring you the best and the brightest authors in the AutoCAD community. Maybe you've read their columns in a CAD journal, maybe you've heard them speak at an Autodesk event, or maybe you're new to these authors—whatever the case may be, we know you'll enjoy and apply what you'll learn from them. We thank you for selecting this title and wish you well on your programming journey.

Sandy Clark
Executive Editor
Autodesk Press

A WORD FROM AUTODESK, INC.

From the birth of AutoCAD onward, there has been a large library of source material on how to use the software; however, relatively little material on customization of AutoCAD has been made available. There have been a few texts written about AutoLISP™, and many general AutoCAD books include a chapter or two on customization. On the whole, however, the number of texts on programming AutoCAD has been inadequate given the amount of open technology, the number of application programming interfaces (APIs), and the sheer volume of opportunity to program the AutoCAD design system.

Four years ago, when I started working with developers as the product manager for AutoCAD APIs, I immediately found an enormous demand for supporting texts about ObjectARX™, AutoLISP, Visual Basic™ for Applications, and AutoCAD OEM. The demand for a "programming series" of books stems from AutoCAD's history of being the most programmable, customizable, and extensible design system on the market. By one measure, over 70 percent of the AutoCAD customer base "programs" AutoCAD using Visual LISP, VBA, or menu customization, all in an effort to increase productivity. The demand for increasing the designer's productivity extends deeply, creating a demand for new source texts to increase productivity of the programming and customization process itself.

The standard Autodesk documentation for AutoCAD provides the original source of technical material on the AutoCAD API technology. However, it tends toward the clinical, which is a natural result of describing the software while it is being created in the software development lab. Documentation in fully applied depth and breadth is only completed through the collective experiences of hundreds of thousands of developers, customers, and users as they interpret and apply the system in ways specific to their needs.

As a result, demand is high for other interpretations of how to use AutoCAD APIs. This kind of instruction develops the technique required to innovate. It develops programmer instinct by instructing when to use one interface over another and provides direction for interpretive nuances that can only be developed through experience. AutoCAD customers and developers look for shortcuts to learning and for alternative reference material. Customers and developers alike want to accelerate their programming learning experience, thereby shortening the time needed to become expert and enabling them to focus sooner and better on their own specific customization or development projects. Completing a customization or development project sooner, faster, and better means greater productivity during the development project and more rapid deployment of the result.

For the CAD manager, increasing productivity through accelerated learning means increasing his or her CAD department's productivity. For the professional developer, this means bringing applications to market faster and remaining competitive.

THE INITIAL IDEA

An incident that occurred at Autodesk University in Los Angeles in the fall of 1997 illustrated dramatically a dynamic demand for AutoCAD API technology information. Bill Kramer was presenting an overview session on AutoCAD's ObjectARX. The Autodesk University officials had planned on having 20 to 30 registrants for this session and had assigned an appropriately sized room to the session. About three weeks before it took place, I received a telephone call indicating that the registrations for the session had reached the capacity of the room, and we would be moving the session. We moved the session two more times due to increasing registration from customers, CAD Managers, designers, corporate design managers, and even developers attending the Autodesk University customer event. What they all had in common was an interest in seeing how ObjectARX, an AutoCAD API, was going to increase their own or their department's design productivity. When the presentation started, I walked into the back of the room to see what appeared to me to be over 250 people in a room that was now standing room only. That may have been the moment when I decided to act, or it may have been just when the actual intensity of this demand became apparent to me; I'm not sure which.

The audience was eager to hear what Bill Kramer was going to say about the power of ObjectARX. Nearly an hour of unexpected follow-up questions and answers followed. Bill had successfully evangelized a technical subject to an audience spanning non-technical to technical individuals. This was a revelation for me, and the beginning of my interest in developing new ways to communicate to more people the technical aspects, power, and benefits of AutoCAD technology and APIs.

As a result, I asked Bill if he would write a book on the very topic he just presented. I'm happy to say that Bill's book has become one of the first books written in the new AutoCAD Programmer Series with Autodesk Press.

TOPICS FOR THE PROGRAMMER SERIES

The extent of AutoCAD's open programmable design system made the need for a series of books apparent early on. My team, under the leadership of Cynde Hargrave, Senior AutoCAD Marketing Manager, began working with Autodesk Press to develop this series.

It was an exciting project, with no shortage of interested authors covering a range of topics, from AutoCAD's open kernel in ObjectARX to the Windows standard for application programming in VBA. The result is a complete library of references in Autodesk's Programmer Series covering ObjectARX, Visual LISP, AutoLISP, customizing AutoCAD through ActiveX Automation™ and Microsoft Visual Basic for Applications, AutoCAD database connectivity, and general customization of AutoCAD.

WHO READS PROGRAMMING BOOKS?

Every AutoCAD user will find books in this series to fit their AutoCAD customization or development interests. The collective goal we had with our team and Autodesk Press for developing this series, identifying titles, and matching them with authors was to provide a broad spectrum of coverage across a wide variety of customization content and a wide range of reader interest and experience.

Collaborating as a team, Autodesk Press and Autodesk developed a programmer series covering all the important APIs and customization topics. In addition, the series provides information that spans use and experience levels from the novice just starting to customize AutoCAD to the professional programmer or developer looking for another interpretive reference to increase his or her experience in developing powerful applications for AutoCAD.

THIRST, THEME, AND VARIATION

I compare this thirst for knowledge with the interest musicians have in listening to music performed in different ways. For me, it is to hear Vivaldi's *The Four Seasons* time and time again. Musicians play from the same notes written on the page, with the identical crescendos and decrescendos and other instructions describing the "technical" aspects of the music.

All of the information to play the piece is there. However, the true creative design and beauty only manifest themselves through the collection of individual musicians, each applying a unique experience and interpretation based on all that he or she has learned before from other mentors, in addition to his or her own practice in playing the written notes.

By learning from other interpretations of technically identical music, musicians benefit the most from a new, and unique, interpretation and individual perception. This makes it possible for musicians to amplify their own experience with the technical content in the music. The result is another unique understanding and personalized interpretation of the music.

Similar to musical interpretation is the learning, mentoring, creative processes, and resources required in developing great software, programming AutoCAD applications, and customizing the AutoCAD design system. This process results in books such as Autodesk's Programmer Series, written by industry and AutoCAD experts who truly love working with AutoCAD and personalizing their work through development and customization experience. These authors, through this programmer series, evangelize others, enabling them to gain from their own experiences. For us, the readers, we gain the benefit from their interpretation, and obtain the value through different presentation of the technical information, by this wide spectrum of authors.

Andrew Stein
Senior Manager
Autodesk Business Research, Analysis and Planning

PREFACE

WHY CONNECT AUTOCAD® TO DATABASES?

Can you remember the first time you used AutoCAD? The day you put down your pencil and began exploring the electronic alternative? Back in the early days when AutoCAD was making its way into the quickly growing PC market, it functioned as a simple computer replacement to the drafting board. With a few clicks of the mouse, you could create primitive objects such as lines, arcs, and text. Rather than using a T square and a pencil, we creatively assembled these primitive objects into drawings that represented something that was to be built in the real world. Even though the drafting was accomplished through a computer, we were still caught up in the paper mentality—we considered the plotted page as the end result. It was not obvious that Computer Aided Design (CAD) had the potential to go far beyond being just an electronic drafting system.

These days, however, the industry is beginning to see the real benefits of using a computer to create intelligent models that more closely represent the physical world around us. Instead of merely producing paper output, our focus has shifted to creating intelligent electronic documents that can provide information to users throughout a project's life cycle.

One way to make a drawing intelligent is to extend the graphic database (the drawing) by linking objects to rows in an external database table. For example, a line object in AutoCAD has inherent graphical properties, such as the *X*, *Y*, and *Z* coordinates of its endpoints. But to the user of the graphical database, a line is not just a line; it also represents something in the real world. If the line represents a sewer pipe, for example, you could link the line object to an external database that provides additional nongraphic information such as material, diameter, and year installed.

The information in the database can also be used to find or select specific objects in the drawing, and it can be used to graphically display the results of database queries. Using an external database to link additional information to AutoCAD objects allows you to use the data as if it were a part of the drawing, while allowing the data to be manipulated outside AutoCAD.

PURPOSE OF THIS BOOK

If you are serious about using databases with AutoCAD, you should start thinking in terms of creating a database application. Even if you don't plan to write any custom programs, successfully linking AutoCAD to a database system requires just as much knowledge of how databases work and warrants just as much up-front planning and design as any custom application. This book will give you the complete picture of AutoCAD database connectivity, so you walk away with the fundamental knowledge you need to link up your databases and begin creating robust database applications with AutoCAD.

WHO SHOULD READ THIS BOOK

This book will be valuable to anyone who uses AutoCAD. As design professionals, we communicate our ideas as we design, create detailed construction documents in order to build, and manage our built environment—all with the help of AutoCAD. Throughout this cycle, there are countless opportunities to go beyond AutoCAD's graphic database and build intelligent models that integrate graphic and nongraphic information. Relational databases are the perfect mechanism for this.

If you are using AutoCAD to manage information during the design process, databases can help you track product information and create a bill of materials. During the construction phase, databases can help manage product vendors, subcontractors, and schedules. If you are a facilities manager, databases can store valuable information about buildings, assets, people, and space utilization. If you manage utility infrastructure, land parcels, or any other geographic information, database connectivity represents one of the most invaluable features of AutoCAD.

If you are a programmer who creates applications for AutoCAD that use external databases, this book will also be valuable to you. The last four chapters cover application design, as well as the programming interfaces needed to create robust database applications. There are also several sample applications and utilities included on the CD in the back of this book.

WHAT YOU NEED FOR THIS BOOK

This book is written for AutoCAD 2000. Since so much has changed in this release of AutoCAD with respect to database connectivity, very little content in this book applies to any earlier version of AutoCAD. If you are familiar with the database connectivity features of earlier releases of AutoCAD, you may find it helpful to refer to Appendix B, which describes changes in functionality, programmability, and vocabulary found in AutoCAD 2000.

To ensure that dbConnect works properly, you should install the version of Microsoft® Data Access Components (MDAC) that ships with AutoCAD 2000. Install MDAC by running MDAC_TYP.EXE from the \data directory on the AutoCAD 2000 installation CD.

You may also find it helpful to obtain the latest revision of MDAC. This includes the latest ODBC drivers, OLE DB Providers, and the latest version of the ActiveX® Data Object (ADO) programming interface. A downloadable install file for MDAC is available on Microsoft's Web site at the following URL:

http://www.microsoft.com/data

HOW THIS BOOK IS ORGANIZED

This book is divided into three sections:

> **The DbConnect User Interface**(Chapters 1 and 2)—These chapters introduce the dbConnect user interface and include several tutorials to help get you started.
>
> **Database Design and SQL** (Chapters 3 and 4)—These chapters explain basic database concepts and show you how to design a database. You will also become familiar with SQL, the language used to communicate with databases.
>
> **Application Development** (Chapters 5, 6, 7, and 8)—These chapters show you how to design and develop custom database applications within AutoCAD and cover the application programming interfaces (APIs) that allow you to communicate with and link to databases.

At the end of the book, you will find the following additional sections:

> **Glossary**—A comprehensive glossary of technical terms used throughout the book.
>
> **DbConnect Command Reference** (Appendix A)—A complete list of AutoCAD commands that invoke the various components of the dbConnect user interface.
>
> **Changes from ASE** (Appendix B)—A summary of the differences between the dbConnect feature of AutoCAD 2000 and the AutoCAD SQL Extension (ASE) feature found in earlier versions of AutoCAD.

WHAT'S NOT COVERED AND WHY

This book does not get into much detail about the nuances of specific database products. There are just too many potential database systems, and many subtle differences, to cover effectively in this book. This book focuses on the technology built into AutoCAD, which is designed to be database-independent. The tutorials and sample applications in the book use Microsoft Access, mainly because an MDB file can easily be included on the CD. The SQL statements in Chapter 4 were written for Access, but they were also tested with SQL Server and Oracle.

HOW TO USE THIS BOOK

This book takes you through the topic of database connectivity from the very basic to the very advanced. You will get the most out of this book if you work through the chapters in order, as most chapters build upon information presented in earlier chapters. There are thirty tutorials in this book. For most of the chapters, especially the first two, you will probably want to be sitting at your computer so you can work through these tutorials. At the end of each chapter, you are encouraged to work through the review questions and exercises as well. Answers to these review questions and exercises can be found at **www.autodeskpress.com**.

CONVENTIONS USED IN THE BOOK

TUTORIALS

The majority of user input for the tutorials is through menus, toolbar icons, and dialog boxes. When a tutorial makes reference to one of these components, it is shown with a **bold sans-serif font**. For example:

> From the **dbConnect** menu, choose **Data Sources**, and then **Connect**.

CODE SYNTAX

When this book shows the syntax of a VBA function, such as a method for an object, the name of the function or method is shown in a bold, non-italics font. Arguments are shown in bold italics, and optional arguments are shown with square brackets [ ]. For example:

> **GetLinks(*LinkTemplate*, [*ObjectIDs*], [*LinkTypes*])**

EXAMPLE CODE

Several chapters of this book include example code. The programming languages used are VBA, AutoLISP and SQL. Due to printing restrictions, line breaks may have been necessary where code would normally have continued on the same line. The ↵ symbol indicates a line continuation in program code. These line breaks are depicted as follows:

```
    For Each objLink In ltSpaces.DbConnect↵
.GetLinks(ltSpaces)
```

All of the example code can be found on the CD in the back of the book in one form or another. If you have trouble reading or understanding the code due to the formatting, you can view (or even print) the code from within the appropriate development environment. For example, VBA code can be viewed and printed from within AutoCAd's VBA Editor.

FEATURES

Throughout this book, a series of real-world examples is used to demonstrate AutoCAD's database connectivity features. Listed below are those examples:

Parcels—A sample GIS application that tracks information about land parcels, such as ownership and street address. This example is used in Chapters 1 and 2.

Class—A sample conference scheduling application that tracks rooms, classes, speakers, and timeslots. This application is used in Chapters 3 and 4.

Office—A sample facilities management application that tracks users and use of space in a typical office building. This application is used in Chapters 5, 6, 7, and 8.

USING THE CD

All of the drawings, databases, and application code needed to work through the tutorials and examples in this book are included on the CD. You should copy the contents of the CD to a single location on your computer or local network. Be sure to clear the **read only** flag on the files after they are copied, as you will be modifying several of the files.

The following table lists the contents of the CD and gives a short description of each file.

File Name	Description
cao.dvb	AutoCAD VBA project containing functions and macros that demonstrate the use of the Connectivity Automation Objects.
circles.dvb	AutoCAD VBA project containing the code for the circles example presented in Chapter 6.
circles.lsp	AutoLISP source code for the circles example.
circles.mdb	Microsoft Access database for the circles example.
class.dbq	AutoCAD query export file containing all of the SQL examples used in Chapter 4.
class.mdb	The Access database for the class example application used in Chapters 3 and 4.
office.dvb	AutoCAD VBA project containing the code for the office example presented in Chapters 7 and 8.
office.dwg	AutoCAD drawing file used for the office example.

continued

File Name	Description
office.mdb	Access database used for the office example.
parcels.dwg	AutoCAD drawing used for the parcels example presented in Chapters 1 and 2.
parcels.mdb	Access database used for the parcels example.
parcels.xls	Microsoft Excel workbook duplicate of the parcels database.
parcels_r14.dwg	A Release 14 version of parcels.dwg used to demonstrate link conversion in Chapter 2.
parcels2.dwg	AutoCAD drawing used for the parcels example.
parcels2.mdb	Access database used for the parcels example.
parcels3.dwg	AutoCAD drawing used for the parcels example.

WE WANT TO HEAR FROM YOU!

This book is part of Autodesk's Programmer Series. We'd like to receive your feedback on this or any of the other titles in this exciting new series. Please contact us at:

The CADD Team, c/o Autodesk Press

3 Columbia Circle, PO Box 15015, Albany, NY 12212-5015

or visit our web site at **http://www.autodeskpress.com**

ABOUT THE AUTHOR

Scott McFarlane is an Associate at Woolpert LLP. He has more than twenty years of programming experience and has been integrating databases with AutoCAD ever since it became possible. He has written several articles on the subject and has presented papers at a variety of conferences. At Woolpert, Mr. McFarlane acts as an application specialist for various Facilities Management and Geographic Information Systems projects.

A WORD ABOUT RELATED TECHNOLOGY TRENDS

Although this book is focused on the connectivity between AutoCAD drawings and external databases, it is important to understand a couple of very important emerging technologies that will have a serious impact on how graphic and non-graphic data can be integrated. The first is object-oriented CAD. The marriage of CAD and object-oriented technology will enable us to work with graphical objects that not only carry the properties of the real-world objects they represent but behave like them too. The C++ development environment known as ObjectARX™ is the foundation for this capability within AutoCAD. For example, a custom object such as a door will not only look like a door and have properties such as width and height, it will also behave like a door. It will be fully aware of its environment and have the ability to respond to changes in that environment. In reality, most doors exist inside walls, and their swing direction is dependent on relationships with other building objects, such as walls, windows, columns, or even electrical switches. Consequently, an AutoCAD object representing a door should exist only in a wall, and if changes are made to the wall, or the door is moved to another location in the wall, the door should be notified of these changes and react appropriately.

Another important technology to be aware of is object-oriented databases. Like the CAD industry, the database industry has recognized the benefits of object-oriented technology. Object-oriented databases give you the ability to create custom data types (objects) that consist of complex data structures and can even have behavior. As CAD professionals, we need to be aware of this, because traditional relational database systems will soon be capable of storing graphical objects in an efficient manner. What is also intriguing is that relational database systems have other significant advantages that are not typically found in CAD systems. These include referential integrity, record-level locking and client/server architecture, and other features that are being added or improved to support geographic data, such as spatial indexing and long transactions.

ACKNOWLEDGMENTS

Throughout the many years that I have been involved in the AutoCAD community, I have met so many people who have given me inspiration and motivation. I would like to thank Andrew Stein at Autodesk for helping create the opportunity for me to write this book, and the team at Delmar Publishers and Autodesk Press for making it happen. I would especially like to thank Allyson Powell for helping me stay focused.

Special thanks also go to Bill Wittreich, my good friend and technical editor, for all his help during the writing process. His experience in technical writing and editing, not to mention his expertise in the subject, was crucial to the completeness and accuracy of the text and tutorials in this book.

Many thanks to my colleagues at Woolpert, specifically database gurus Ben Kinser and Brian Hinze, for their help with the chapter on SQL, and to Bill Kimbrell for the many in-depth discussions on VBA and Visual LISP.

Last, but not least, I would like to express my sincere thanks to my wife, Lynn, and my son, Jackson, for their patience and tolerance, and for allowing me to put off working on the house for a few months.

The team at Delmar Publishers and Autodesk Press would like to express their sincere thanks to Bill Wittreich for his keen technical editing and his assistance during the development of this text. His professionalism and willingness to go above and beyond were greatly appreciated.

The team would also like to thank the compositor, Vince Potenza of SoundLightMind Media Design & Development, the copyeditor, Gail Taylor, and the proofreader, Harriet J. Hart, for their contributions to this text.

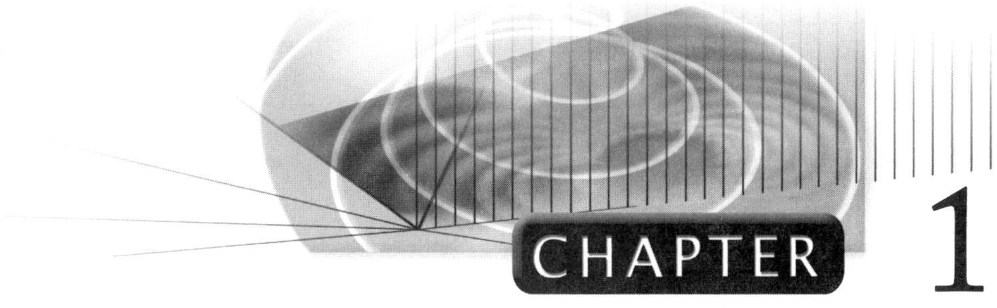

An Introduction to dbConnect

OBJECTIVES

After completing this chapter, you will be able to

- Establish a connection to a database
- View table data using the Data View window
- Link objects in the drawing to a table in the database
- Create labels
- Create simple queries

INTRODUCTION

A BRIEF HISTORY

Ever since Release 12, when Autodesk introduced the AutoCAD SQL Extension™ (ASE), AutoCAD users have had the ability to link AutoCAD objects to external databases. And even prior to Release 12, there were several third-party solutions that accomplished this task. Before these products became available, the only way you could attach non-graphic data to AutoCAD objects was to use blocks and attributes. Attributes are text objects that can be defined along with a block definition. When an instance of a block is inserted, the user is prompted for the text value of each attribute.

Conceptually, the block definition with its attribute definitions could be treated like a simple database table structure, and the multiple instances of the blocks could be treated like rows in a table. In fact, you could export the attribute information to a text file that could be easily imported to an external database system. This capability soon proved to be invaluable in many industries. If you were using AutoCAD to draw real world objects, either for design or for management, you could attach additional non-graphic information to the objects using attributes.

But this approach has many limitations. First, attributes can only be attached to blocks and not to any other type of AutoCAD object. Second and most importantly, the transfer of information from the drawing to the database is a one-way operation. You could not make changes to the extracted data outside AutoCAD and expect those changes to be realized on the attributes. For applications such as facilities management, this was a critical limitation. Most applications that connect graphics to databases, like facilities management, needed to behave more like database applications than graphical applications. The database management needed to be done where it belonged—in a database application, not inside AutoCAD. AutoCAD's role in the whole process was simply to graphically represent the real world objects being managed in the database application.

ASE opened the door to connecting AutoCAD to external databases. Now the data associated with an application could be managed in a database environment, and AutoCAD objects could be linked to the database. ASE used SQL (Structured Query Language), which is an industry-standard language used to talk to databases. This made it easy for someone familiar with databases to get up and running quickly. The early versions of ASE (in AutoCAD for DOS) used special drivers to allow AutoCAD to link to a variety of databases. Then AutoCAD for Windows introduced support for Microsoft's® ODBC (open database connectivity), which provided connections to any database that had an ODBC driver.

In AutoCAD 2000, the database connectivity features are based on Microsoft's OLE DB technology. OLE DB is conceptually similar to ODBC in that it acts as a communication layer between an application and the database management system. However, OLE DB has been designed to make it easier to work with databases that are not necessarily stored in a relational form. It also addresses some of the challenges that the Internet has imposed on database connectivity, such as distributed databases. While ODBC is still widely used, OLE DB may eventually replace ODBC as the new standard for universal data access.

Another key difference is that OLE DB can be completely controlled through a Component Object Model (COM) programming interface known as ActiveX® Data Objects (ADO). ADO can be used from any computer language that supports COM. This includes the AutoCAD programming environments, such as Visual Basic for Applications (VBA) and Visual LISP™. Chapter 6 examines ADO in more detail and demonstrates how to use ADO in these environments.

EXAMPLE APPLICATIONS

Throughout this book, a series of real world examples is used to demonstrate AutoCAD's database connectivity features. Listed below are the examples:

Parcels—Tracking information about land parcels

Class—A sample conference scheduling application that tracks rooms, classes, speakers and timeslots

Office—Managing users and use of space, as well as assets in a typical office building

GETTING STARTED

DBCONNECT AND ASE

AutoCAD 2000 combines all of its database connectivity features in a single, comprehensive user interface called *dbConnect*. DbConnect completely replaces the old AutoCAD SQL Extension (ASE) interface found in earlier releases of AutoCAD. For those of us who struggled with ASE, dbConnect is a welcome change.

Note: Veteran AutoCAD users who previously used ASE should be aware that dbConnect represents a total replacement of ASE. This includes all ASE commands, dialog boxes and AutoLISP functions. ASI drivers are also no longer used. Don't worry, all of the features found in ASE have been included in dbConnect. For more information on command and terminology changes from ASE, see Appendix B.

BASIC DBCONNECT FEATURES

The following are just a few of the important features of dbConnect:

- Select objects using queries
- Create text objects that are based on linked data values
- View and edit table data in an easy-to-use spreadsheet-like interface
- Create new database tables from a selection of linked objects

USER INTERFACE COMPONENTS

The dbConnect graphical user interface (GUI) is made up of the following components:

- The Data Source Configuration utility
- The dbConnect Manager
- The Data View window
- The Query Editor

INVOKING DBCONNECT COMMANDS

There are three basic methods you can use to invoke the various commands associated with dbConnect: the toolbars and shortcut menus within dbConnect Manager, the pull-down menus and, of course, the command line. As you work through the tutorials in this chapter, you will probably find the toolbars and menus to be the most efficient methods. However, there are times when a repetitive task can be accomplished more quickly through the command line. If you are a die-hard command line fan, there is a complete list of dbConnect commands in Appendix A.

THE DBCONNECT MANAGER

The dbConnect Manager is the primary interface to AutoCAD's database connectivity features. It is a dockable window that has a tree view and a tool bar. The tree view has one root node for each open drawing and a single root node that expands into the list of available *data sources* (described below). Once a connection to a data source has been established, its node can be expanded to display the list of the tables in the data source. Many dbConnect commands can be invoked from the tree view through shortcut menus. Display shortcut menus by clicking your right mouse button on items in the tree view.

The dbConnect Manager window can be toggled on and off through any one of the following methods:

- Click the **dbConnect** icon on the standard toolbar (Figure 1.1)
- Select **dbConnect** from the **Tools** menu (Figure 1.2)
- Press CTRL + 6
- Type "dbconnect" or "dbcclose" at the command line

Figure 1.1 *dbConnect on the Standard Toolbar*

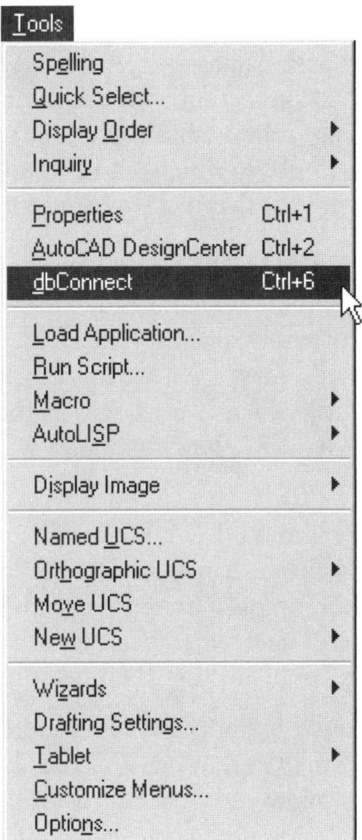

Figure 1.2 *dbConnect on the Tools Menu*

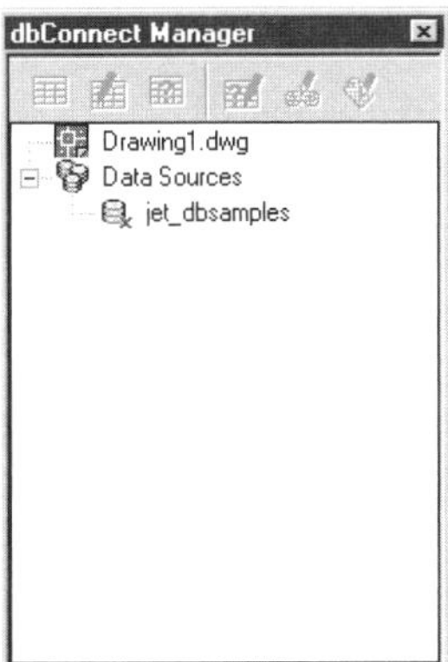

Figure 1.3 *The dbConnect Manager*

ESTABLISHING A DATABASE CONNECTION

Before you can do anything with databases in AutoCAD (or any application for that matter), you must establish a connection with the database. Most database systems employ some level of security, and by establishing a connection you are, in essence, logging in to the database system as a specific user. In AutoCAD 2000, you must define a *data source* for any database to which you are connecting from within AutoCAD.

DATA LINKS AND DATA SOURCES

A data source contains all the information needed to establish a connection with a particular database system. The amount of information needed to create a data source depends on the type of database system you are connecting to. In this chapter, you will be using the Parcels example, which is included on the CD. This example uses a Microsoft Access database. Chapter 2 shows additional examples of connecting to other database systems.

Since AutoCAD 2000 is using the OLE DB technology, the data source configuration is handled by a utility that is a part of Microsoft's data access components. OLE DB uses the term *data link* to describe a specific database connection, while AutoCAD 2000 uses the term *data source*. You may see these two terms used interchangeably throughout this book, but they both describe the same concept.

A Microsoft data link is similar to an ODBC data source, except that data links are stored as files on disk rather than in the Registry. You can create a data link file from within Explorer by selecting **New** and then **Microsoft Data Link**. This creates a file with a *.udl* extension that contains all the connection information for the data source.

AutoCAD 2000 uses the same UDL files to connect to external databases. AutoCAD looks for the UDL files in a configurable directory location. By default, UDL files are stored in a subdirectory of the AutoCAD 2000 installation directory called "Data Links." Rather than creating new data links from within Explorer, you can create them from within AutoCAD, and they are automatically stored in the appropriate directory. In AutoCAD, this is called *configuring a data source.*

The following tutorial takes you through the steps required to configure a data source for the Parcels example database.

TUTORIAL 1.1 – CONFIGURING A DATA SOURCE

1. Start AutoCAD 2000.
2. From the **Tools** menu, choose **dbConnect**. This displays the dbConnect Manager shown in Figure 1.3. A **dbConnect** pull-down menu is also added to the AutoCAD 2000 menu bar.

Tip: You can also toggle the display of the dbConnect Manager using the **dbConnect** icon on the standard toolbar, or by typing CTRL + 6.

3. Right-click the Data Sources branch in the dbConnect Manager and choose **Configure Data Source...** from the shortcut menu. The **Configure a Data Source** dialog box is displayed, as shown in Figure 1.4.

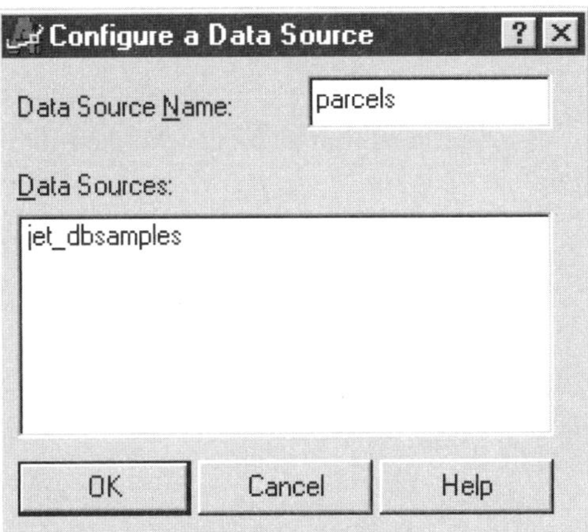

Figure 1.4 *The Configure a Data Source Dialog Box*

4. Type "parcels" as the data source name, and click **OK** or press ENTER. This launches the **Data Link Properties** dialog box, as shown in Figure 1.5.

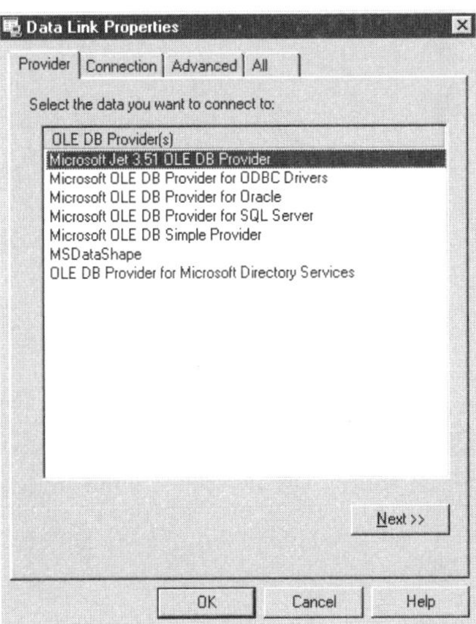

Figure 1.5 *The Data Link Properties Dialog Box*

5. Select **Microsoft Jet 3.51 OLE DB Provider** in the list of OLE DB Providers, and click **Next >>**.

Note: If you do not see the Microsoft Jet 3.51 OLE DB Provider in the list of providers, you may need to install the version of Microsoft Data Access Components (MDAC) that ships with AutoCAD 2000. Install MDAC 2.0 by running MDAC_TYP.EXE from the \data directory on the AutoCAD 2000 installation CD.

6. Enter the full path of the parcels.mdb file. This file is included on the CD-ROM.

Tip: You can use the **ellipsis** icon located just to the right of the edit box to bring up a file dialog box.

7. Click **Test Connection** to verify that a connection can be made.
8. Click **OK**.

Once a data source has been configured, a UDL file is created in the Data Links directory, and its name appears under the Data Sources node in the dbConnect Manager window as shown in Figure 1.6.

Figure 1.6 *Data Sources Node in the dbConnect Manager*

Note: The dbConnect interface does not provide a way to delete a data source. If you need to delete a data source, you must delete the associated UDL file.

CONNECTING TO THE DATABASE

You can establish a connection to a database using one of the following methods:

- Double-click the data source name in the dbConnect Manager
- Right-click the data source name in the dbConnect Manager and choose **Connect**
- From the **dbConnect** menu, choose **Data Sources**, and then **Connect**

Note: Data sources that are not connected have a small red "X" beside the icon in the dbConnect Manager. When you make a successful connection, the "X" is removed.

VIEWING AND EDITING TABLE DATA

AutoCAD 2000 provides a built-in spreadsheet-like interface for viewing and editing table data called the *Data View* window. The Data View window functions much like the table-viewing window in Microsoft Access. You can move and resize columns, freeze columns, sort, and search and replace data. The Data View window is the primary interface to working with tables that are linked to AutoCAD objects.

In the following section, we will examine the different parts of the Data View window using the Parcels example. This is only intended to be an overview. Chapter 2 includes a more detailed discussion of the Data View window and its capabilities.

If you want to follow along, you can connect to the parcels data source using one of the methods described above. The parcels node in the tree view will expand to reveal the available tables in the database. Double-click the **PARCEL_DATA** table. This launches the Data View window, as shown in Figure 1.7. Also notice that a **Data View** pull-down menu (Figure 1.8) now appears on the AutoCAD menu bar.

Data View - PARCEL_DATA (Drawing1.dwg)

-- new Link Template -- | -- new Label Template --

PARCEL_ID	OWNER_NAME	ADDRESS	STREET	ACRES	LAND_VALUE
2945-12-42-01	Ahlbrand, Stephen & Ferrell		Big Pine Rd	0	36696
2945-12-42-02	Alyanak, Edward J & Deborah Jones		Big Pine Rd	0	20453
2945-12-42-03	Atkinson, David E & Lana M		Big Pine Rd	0	30867
2945-12-42-04	Bare, Kristopher R & Rebecca		Big Pine Rd	0	38323
2945-12-42-05	Barnett, Wendell D & Janet B		Big Pine Rd	0	28605
2945-12-42-06	Baxter, Todd A & Karen L		Big Pine Rd	0	33559
2945-12-42-07	Bebo, Rodger A & Nancy T		Big Pine Rd	0	30049
2945-12-42-08	Boddie, Charles E		Big Pine Rd	0	30275
2945-12-42-09	Brands, Michael C & Janice M		Big Pine Rd	0	29260
2945-12-42-10	Brueggemann, Stephen G & Lisa A		Divorce Ct	0	27069
2945-12-42-11	Buchanan, Dennis A & Loretta L		Divorce Ct	0	28097
2945-12-42-12	Burnett, Paul & Jane		Divorce Ct	0	25395
2945-12-42-13	Calhoun, Lee W		Divorce Ct	0	21112

Record 1

Figure 1.7 *The Data View Window*

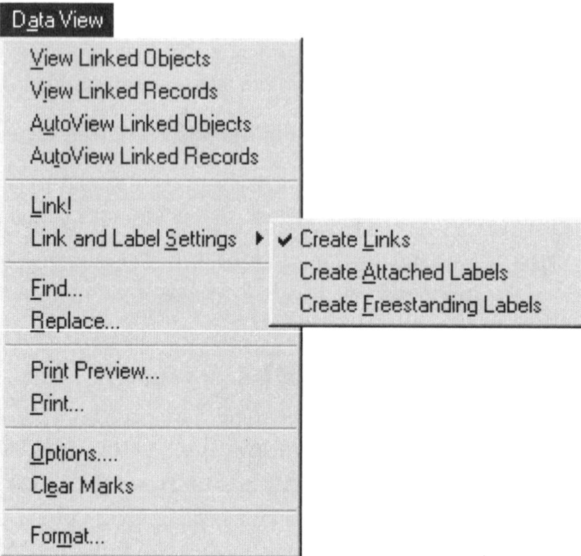

Figure 1.8 *The Data View Pull-down Menu*

At the top of the Data View window is a toolbar that gives you quick access to the most frequently used functions. Some of the buttons may be grayed out since we don't have any objects linked yet.

Just below the toolbar is the table data in a "grid" format. Each column in the grid has a column header that displays the column name as it is stored in the table. Column selection is done through standard Windows conventions. To select a column, simply click the column header. To select multiple columns, drag the mouse across the headers you want to select or hold down SHIFT to select adjacent columns or CTRL to select non-adjacent columns.

Using your mouse, you can double-click a column header to quickly sort the table based on the data in that column. Double-click again in the same column header to reverse the sort order. You can resize the display width of a column by dragging the dividing line between that column and the column immediately to its right. You can hide a column by simply dragging its width to the left until the column header disappears. You can also rearrange the columns by positioning the mouse pointer in the middle of the column header and dragging the column to the desired position.

To view the column shortcut menu, position the mouse pointer in the middle of any column header and right-click. Choose **Sort** on the shortcut menu to display the **Sort** dialog box. This dialog box is useful if you need to sort the table data by more than one column. For example, you could sort the data in the parcel table first by street name, and then by owner's name.

Also on the shortcut menu is a **Hide** option, which temporarily hides any selected columns. To redisplay any hidden columns, choose **Unhide All** from the shortcut menu.

From this shortcut menu, you can also **Freeze** one or more columns. Freezing columns locks them in place at the left side of the Data View window so that they always stay visible, even when you scroll horizontally to the right.

LINKING OBJECTS TO THE DATABASE

The ability to link objects to specific rows in an external database table is the essence of AutoCAD's database connectivity feature. In the Parcels example, we have a drawing that graphically represents a map of the parcels and a table in our database that contains one row for each parcel. The next step is to create the link between the parcels on the map and the corresponding rows in the table. It is important to note that each parcel in the table has a unique ID (in the **PARCEL_ID** column). In order to link an object to a row in the table, you must be able to uniquely identify that row from all other rows in the table. You can accomplish this using a single **ID** column (as it is done in the Parcels example) or multiple columns, if necessary. This column (or columns) is generally referred to as the *key* or *primary key* column.

A link is simply a piece of information stored on an AutoCAD object that can uniquely identify a row in an external table. Any type of AutoCAD object can have one or more links. The information needed to establish a link between an object and a row in the table is as follows:

Data Source—The database management system used

Catalog—The database; a data source may contain one or more catalogs (databases)

Schema—The subset of the database that is available to a particular user

Table—The table

Key column(s)—The names of the columns that uniquely identify the row in the table

Key value(s)—The specific values that are used to find the row that is being linked

Notice that this list represents a hierarchy: A data source contains multiple catalogs, a catalog contains multiple schemas, a schema contains multiple tables, and a table contains multiple columns. This hierarchy comes from the SQL standard upon which dbConnect is based. SQL is discussed in more detail in Chapter 4.

Typically, you may have hundreds or even thousands of objects linked to different rows in the *same* table. This means that for each object, the data source, catalog, schema, table, and key column are the same. The only thing different on each object is the key value. As you can imagine, this would result in a tremendous amount of unnecessary duplication. To eliminate this duplication, AutoCAD 2000 uses a concept called the *link template*. A link template stores everything about a particular link except the link value. Then for each object you want to link, you simply specify the link template and then the specific key column values for that object. Another advantage to this is that it gives you the flexibility to change the database system you are using at any time, without having to relink the objects. Just change the data source name in the link template.

CREATING LINK TEMPLATES

You must create a link template before you can start creating links on entities. Since link templates are associated with a particular table, you must be connected to a data source before they can be created. There are a number of ways to create a link template within the dbConnect user interface. In the dbConnect Manager, you can right-click a table name and choose **New Link Template** from the shortcut menu or use the toolbar button. You can also use the **dbConnect** menu.

TUTORIAL 1.2 – CREATING LINKS TO GRAPHICAL OBJECTS

1. Open parcels.dwg. This drawing contains four closed polygons that, for the purposes of this tutorial, represent parcel boundaries.
2. Open the dbConnect Manager, and connect to the parcels data source.
3. Right-click the **PARCEL_DATA** table and choose **New Link Template**. The **New Link Template** dialog box is displayed, as shown in Figure 1.9.
4. Accept the default link template name of **PARCEL_DATALink1** by clicking **Continue**. This takes you to the **Link Template** dialog box (Figure 1.10), where you indicate which columns in the table you are using as your link columns.
5. Click the box next to **PARCEL_ID** to indicate the key column you are using, and then click **OK**. You should see a **PARCEL_DATALink1** node in the dbConnect Manager just below parcels.dwg.
6. Before you start creating links on entities, you must bring up the Data View window. Right-click the **PARCEL_DATALink1** node and choose **View Table**.

Tip: You may find it easier to "dock" the Data View window so that it does not obscure your drawing area. To ensure that docking is enabled for the Data View window, right-click the Data View window toolbar and choose **Allow Docking**. Then drag the window by its title bar to one edge of the AutoCAD window.

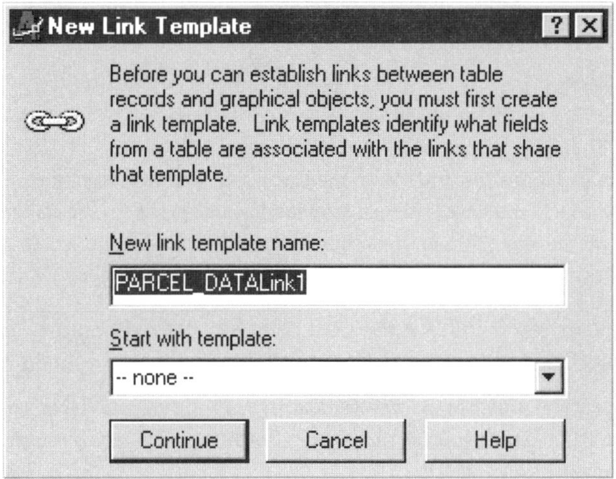

Figure 1.9 *The New Link Template Dialog Box*

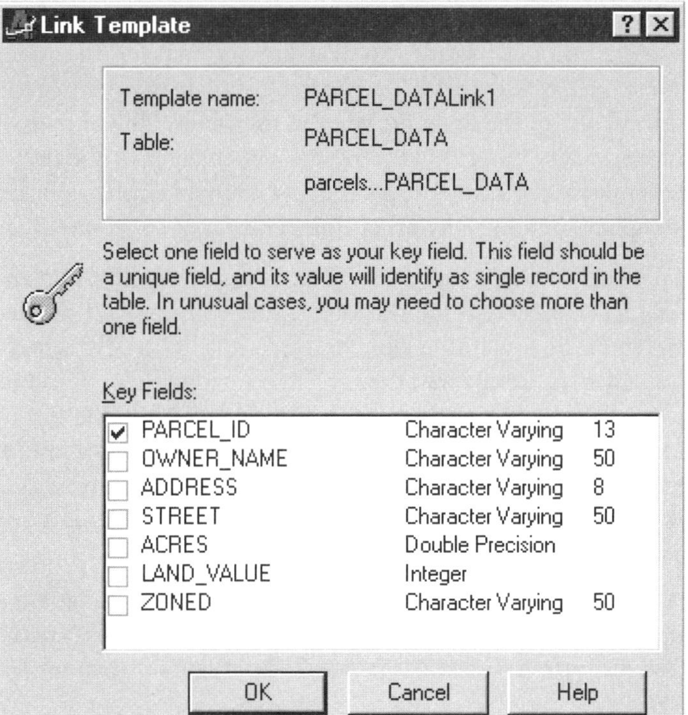

Figure 1.10 *The Link Template Dialog Box*

7. In the Data View window, click the row selector (just to the left of the row) for the first row in the table.
8. Click the **Link** toolbar button and select the leftmost parcel boundary.
9. Press ENTER.

Note: Notice that the row in the Data View window is highlighted in yellow and that the next row in the table has been selected. If your objects on the screen are in an order similar to the way they appear in the table, this feature allows you to continue linking rows without having to select the next row in the Data View window each time.

10. Repeat step 8 and link the next parcel to the right to the next row in the table. Continue to the right until all four parcels have been linked.

Tip: If you prefer to use the keyboard, you can use the DVLINK command to create the first link, and then you can simply press SPACEBAR or ENTER to repeat the command as you continue linking objects.

11. Save the drawing.

CREATING LABELS

Labels are text objects in AutoCAD that display information extracted from the database. For example, you may want to show a drawing that has the street address and owner's name shown as text within each parcel. Rather than duplicating this information as "dumb" text in the drawing, you can use labels. Unlike normal text objects, labels can automatically be updated to reflect any changes to the data. Since a label object displays information from a specific row in a table, it has a link (and an associated link template) just like any other linked object.

DbConnect provides two types of label objects: a *freestanding label* and an *attached label.* A freestanding label is a Multiline text object that contains text derived from actual column values from the row in the table to which it is linked. Attached labels are just like freestanding labels, except that they are associated with an object in the drawing that is linked to the same row. Attached labels include a Multiline text object, a leader object and the associated object. Each of these three objects has a link, and if the link on one object changes, the links on the other objects are automatically changed. If you move the associated object, the leader and label move with it. If you delete the object, the attached label and leader are also deleted.

Tip: At times, you may want to used attached labels, but you do not need the leader objects. While there is no way to eliminate the leader objects when attached labels are used, you can modify the leaders, so that they reside on their own layer, and then keep that layer turned off.

As for linked objects, it is common to have many label objects in a drawing that display the same column values and have the same text formatting and object properties.

Rather than defining this information on each label, AutoCAD uses a concept called the *label template*.

CREATING LABEL TEMPLATES

To create freestanding or attached labels, you must first establish both a link template and a label template. The label template describes how text is displayed in the labels. Since labels are implemented as MTEXT entities, the label template editor is really a modified version of the MTEXT editor. In the following tutorial, you will place a label showing the owner's name inside each parcel.

Note: Since you already have a link template established that describes the link to the **PARCEL_DATA** table, it is not necessary to create a new link template just for the labels.

TUTORIAL 1.3 – CREATING FREESTANDING LABELS

1. Open parcels.dwg, display the dbConnect Manager, and connect to the parcels data source if it is not already connected.
2. Right-click the **PARCEL_DATA** table and choose **New Label Template**. The **New Label Template** dialog box is displayed, as shown in Figure 1.11.

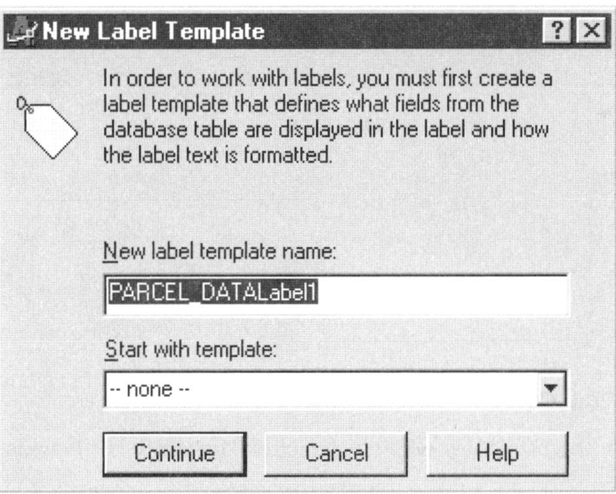

Figure 1.11 *The New Label Template Dialog Box*

3. Accept the default label template name of **PARCEL_DATALabel1** by clicking **Continue**. This takes you to the **Label Template** dialog box as shown in Figure 1.12.

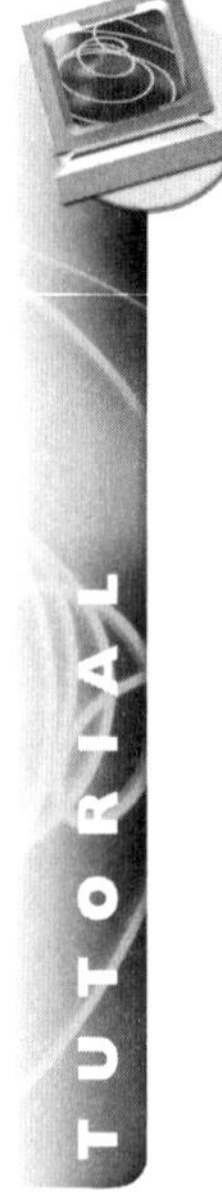

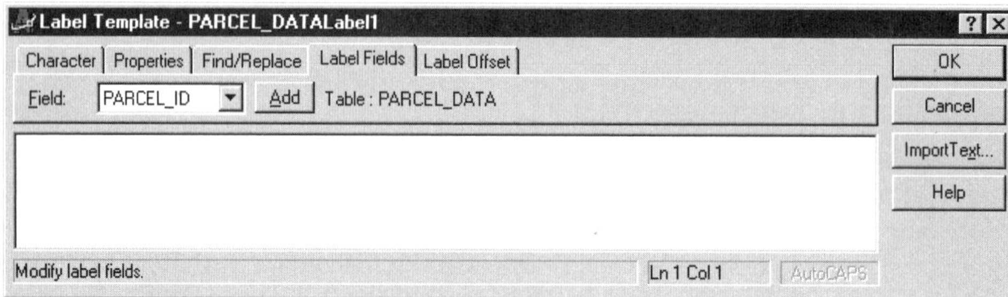

Figure 1.12 *The Label Template Dialog Box*

4. Select the **Character** tab and change the font height to 4.0.
5. Select the **Properties** tab and change the justification to Middle Center.
6. Return to the **Label Fields** tab, select **OWNER_NAME** from the **Field** drop-down list and click **Add**. The **Label Template** dialog box should now look like Figure 1.13.

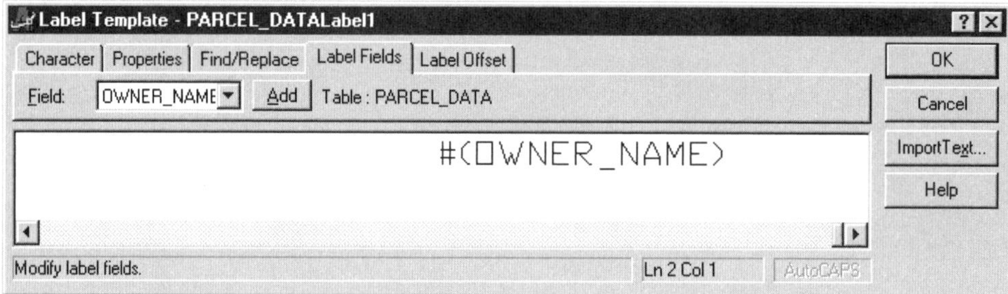

Figure 1.13 *The Label Template Dialog Box*

7. Click **OK**.
8. In the Data View window, select the first row in the table.
9. Next to the **Link** toolbar button there is a small down arrow. Click this arrow and choose **Create Freestanding Labels** (Figure 1.14). Notice that the **Link** button icon changes to a **Label** icon.
10. Click the **Label** toolbar icon. At the command line, you are prompted to select an insertion point for the label. Select a point in the center of the first parcel.

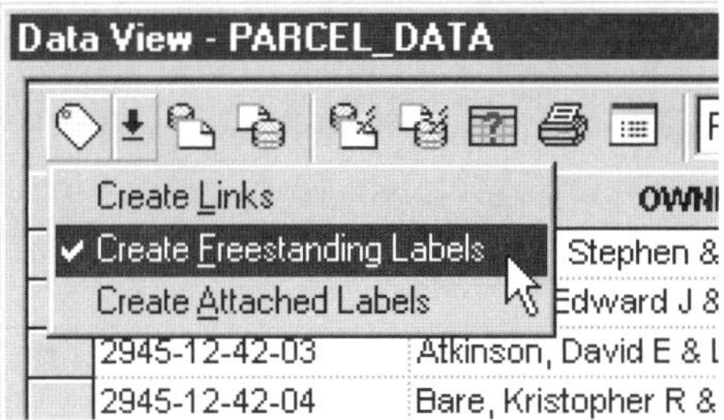

Figure 1.14 *Select the Create Freestanding Labels Tool*

Tip: As with links, you can type "DVLINK" at the command line to create the first label, and then you can simply press SPACEBAR or ENTER to repeat the command as you continue adding labels.

11. Repeat step 10 until you have placed labels in each of the four parcels.
12. Save and close the drawing.

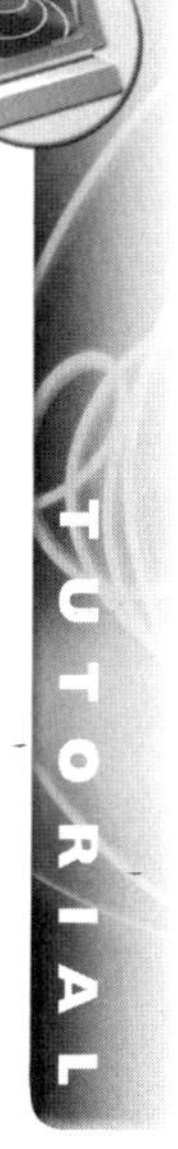

THE QUERY EDITOR

There are times when you want to work with just a subset of the data in your database. To select only the rows you want, you can create a query. Queries are the fundamental mechanism by which information is retrieved from a database. In AutoCAD 2000, queries can be formulated with the dbConnect *Query Editor.* The Query Editor gives you four different ways to create queries:

Quick Query—For very simple queries

Range Query—To query based on a range of values

Query Builder—For more complex queries

SQL Query—For advanced users who just want to type the necessary SQL statements to create the query

The first two methods are covered in this chapter. Chapter 2 demonstrates the latter two methods, and Chapter 4 demonstrates even more powerful examples of using the *SQL Query* method, as we discuss the use of SQL in more detail.

When you execute a query using the dbConnect Query Editor, the resulting rows can be displayed in the Data View window. You also have the option of highlighting the objects in the drawing that are linked to the resulting rows.

Queries can be saved within the drawing for later use. Queries stored in the drawing appear as child nodes of the drawing in the dbConnect Manager. You can also save a drawing's query set to an external file. Query sets saved to a file have a *.dbq* extension.

QUICK QUERY

The simplest form of query is called a Quick Query. The Quick Query can be used if you want to search a table for a single column value. In the following tutorial, you will query the database for parcels that are on Big Pine Rd and highlight the parcels in the drawing.

TUTORIAL 1.4 – CREATING A SIMPLE QUERY

1. Open parcels2.dwg. This drawing is similar to parcels.dwg, except that there are more parcels, and all the parcels have links.
2. Display the dbConnect Manager, and connect to the parcels data source if it is not already connected.
3. Right-click the **PARCEL_DATA** table and choose **New Query** from the shortcut menu. This displays the **New Query** dialog box as shown in Figure 1.15.

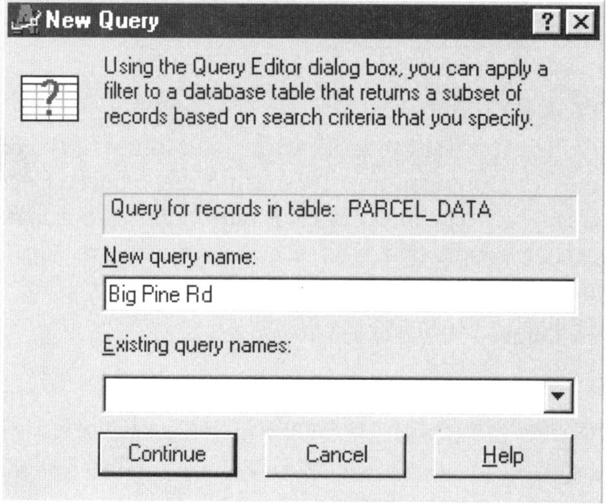

Figure 1.15 *The New Query Dialog Box*

4. Type "Big Pine Rd" as the query name and click **Continue**. This takes you to the **Query Editor**, as shown in Figure 1.16.
5. Make sure the **Quick Query** tab is selected.
6. In the **Field** list select **STREET** and make sure **= Equal** is selected in the **Operator** pick list.
7. Click the **Look up values** button. This displays all the distinct values for the **STREET** column as shown in Figure 1.17.

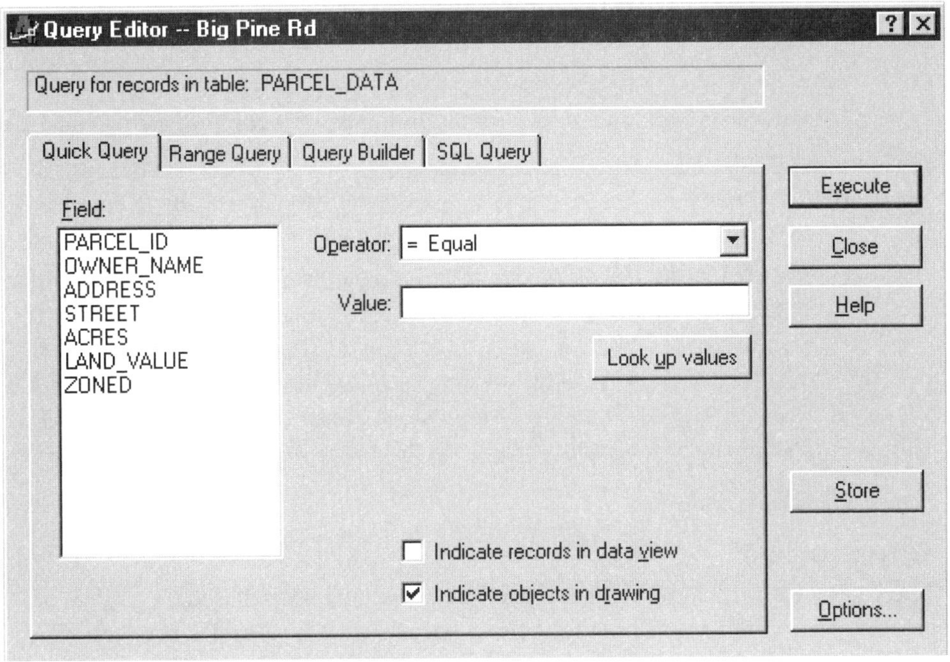

Figure 1.16 *The Query Editor Dialog Box*

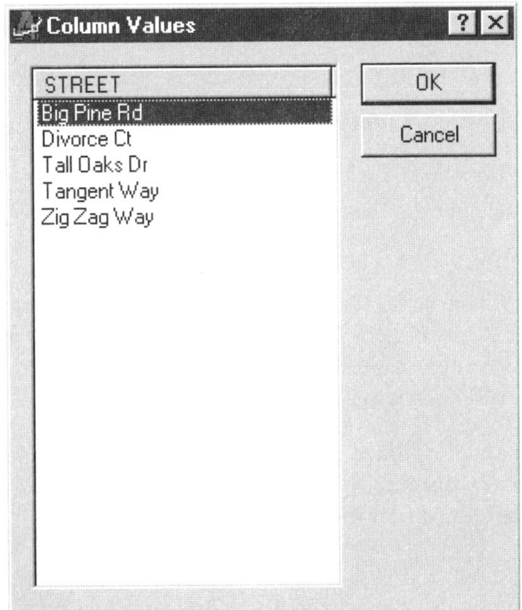

Figure 1.17 *The Column Values Dialog Box*

8. Select **Big Pine Rd** and click **OK**. The text "Big Pine Rd" now appears in the Value field.
9. At the bottom of the dialog box, deselect the **Indicate records in data view** check box, and make sure the **Indicate objects in drawing** check box is selected.
10. Click **Store** to save the query in the current drawing. The query appears as a node in the dbConnect Manager under the parcels2.dwg node.
11. Click **Execute**. The dialog box is closed, and the parcels along Big Pine Rd are highlighted (selected) in the drawing.

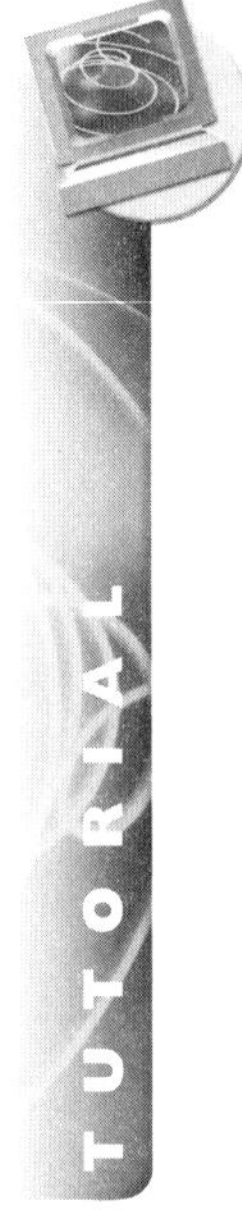

Note: The objects are selected the same way they would be if you had picked them manually. To deselect them, simply press ESC a couple of times. You may also notice that when the dbConnect Manager window is open, you may need to click in the command prompt area or the graphics area to shift input focus so that ESC will work properly.

12. Save the drawing.

RANGE QUERY

The Query Editor also provides an easy way to create a query based on a range of values. Let's say you want to highlight those parcels with land values between $29,000 and $31,000. To accomplish this, follow these steps.

TUTORIAL 1.5 – CREATING A RANGE QUERY

1. Right-click the **PARCEL_DATA** table and choose **New Query** from the shortcut menu to display the **New Query** dialog box.
2. Type "Land Value" as the query name and click **Continue**.
3. Select the **Range Query** tab in the **Query Editor**.
4. In the **Field list**, select **LAND_VALUE**.
5. Type "29000" in the **From** field.
6. Type "31000" in the **Through** field.

Tip: At any time during the query building process, you can see what your query looks like as an SQL statement by selecting the **SQL Query** tab. If you do this, however, make sure you don't make any changes to the SQL statement. If you do, the **Query Editor** will reset itself if you try to go back to any other tab.

7. Make sure the **Indicate records in data view** check box is deselected, and the **Indicate objects in drawing** check box is selected. The **Query Editor** dialog box should now look like Figure 1.18.
8. Click **Store** to save the query in the current drawing.
9. Click **Execute** to dismiss the dialog box and highlight (select) the parcels that satisfy the query.

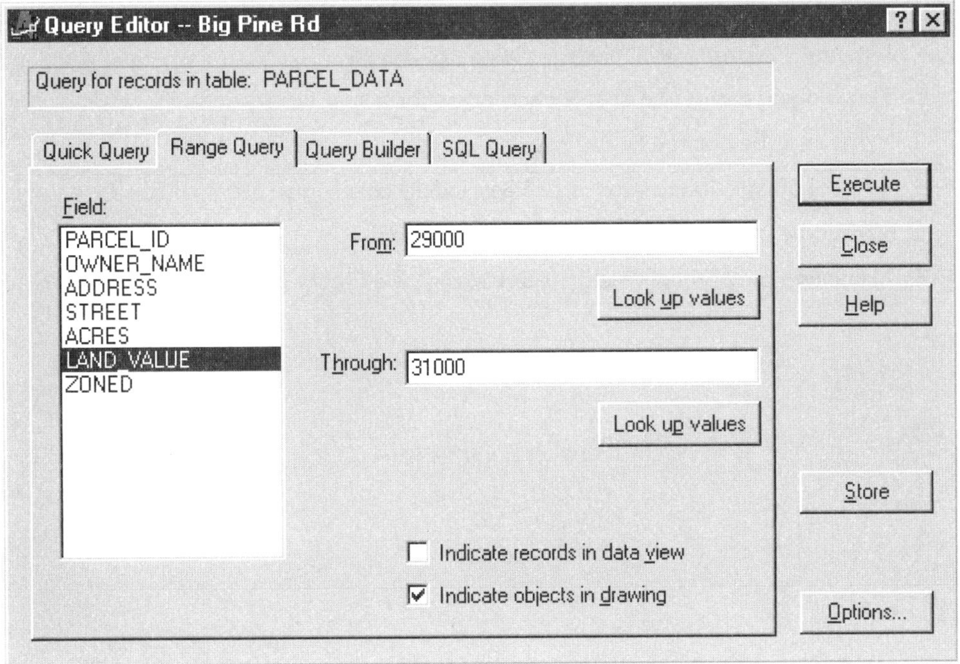

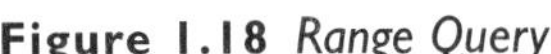

Figure 1.18 *Range Query*

SUMMARY

You now have been introduced to some of the basic features of dbConnect. You have learned how to

- Establish a connection to a database
- View table data using the Data View window
- Link objects in the drawing to a table in the database
- Create labels
- Create simple queries

In the next chapter, you will learn some of the more advanced features of dbConnect.

REVIEW QUESTIONS

1. On what Microsoft technology is dbConnect based?
2. How do you display the Data View window for a specific table?
3. What must be created before you can start linking entities to a table?
4. What is the difference between a freestanding label and an attached label?
5. What type of AutoCAD object is used for labels?
6. Name two methods of creating queries using the Query Editor.

Advanced dbConnect Features

OBJECTIVES

After completing this chapter, you will be able to

- Work with your linked data using the Data View window
- Create more complex queries using the Query Editor
- Build selection sets using Link Select
- Export the links of selected objects
- Modify link template properties
- Use the Synchronize feature to check link integrity
- Convert links from older drawings
- Link to Excel and other database systems

INTRODUCTION

You now have a basic idea of how the various dbConnect interface components are used to work with external databases. This chapter provides more detail on some of the features you have already learned and takes you further into the more advanced dbConnect features.

Throughout this chapter, we will be using the Parcels example drawing and database that we used in Chapter 1.

ADVANCED DATA VIEW FEATURES

In Chapter 1 you learned some of the basic features of the Data View window. The Data View window provides several features that make it a very powerful and flexible interface to your table data.

THE DATA VIEW INTERFACE

Before we get into the specific features, we need to take a closer look at the user interface components of the Data View window. The various commands and other functionality related to the Data View window are accessible through the following user interface components.

- The main **Data View** menu
- The **Data View** toolbar
- Shortcut menus

The Data View Menu

Figure 2.1 shows the **Data View** menu, which can be found on the main AutoCAD menu bar when the Data View window is visible.

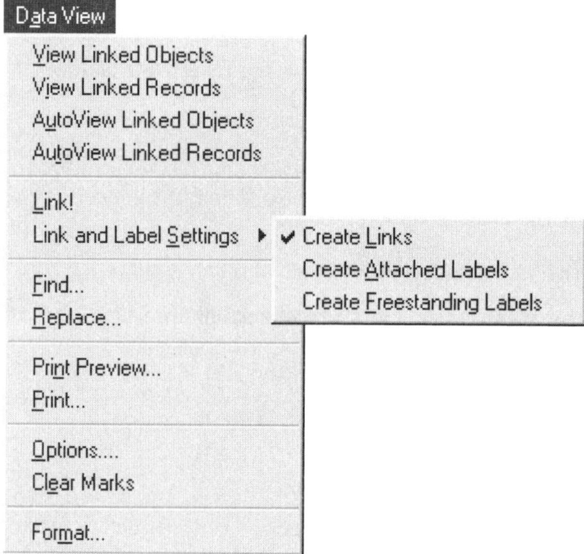

Figure 2.1 *The Data View Menu*

The Data View Toolbar

Figure 2.2 shows the toolbar that appears at the top of the Data View window.

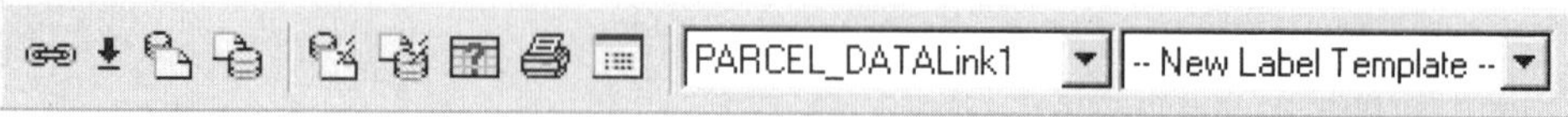

Figure 2.2 *The Data View Toolbar*

The tools and icons on the **Data View** toolbar, from left to right, are as follows:

- Link (or Create Labels)
- Link and Label Settings (down arrow)
- View Linked Objects in Drawing
- View Linked Records in Data View
- AutoView Linked Objects in Drawing
- AutoView Linked Records in Data View
- Query
- Print Data View
- Data View and Query Options
- Link Template Pick List
- Label Template Pick List

Shortcut Menus

Using your right mouse button while your pointer is positioned over some object or area on the screen activates a *shortcut* menu. In the Data View window, there are six distinct shortcut menus that can be displayed, depending on where your mouse pointer is positioned.

Figure 2.3 *The Window Shortcut Menu*

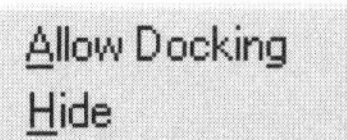

Figure 2.4 *The Toolbar Shortcut Menu*

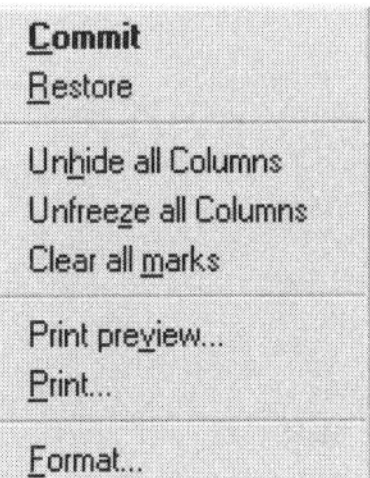

Figure 2.5 *The Table Selector Shortcut Menu*

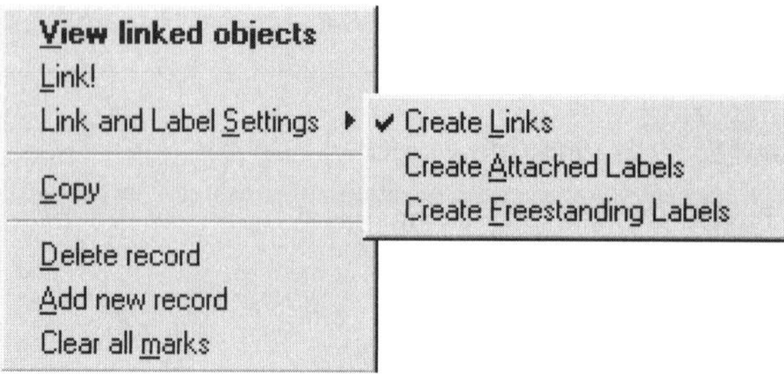

Figure 2.6 *The Row Selector Shortcut Menu*

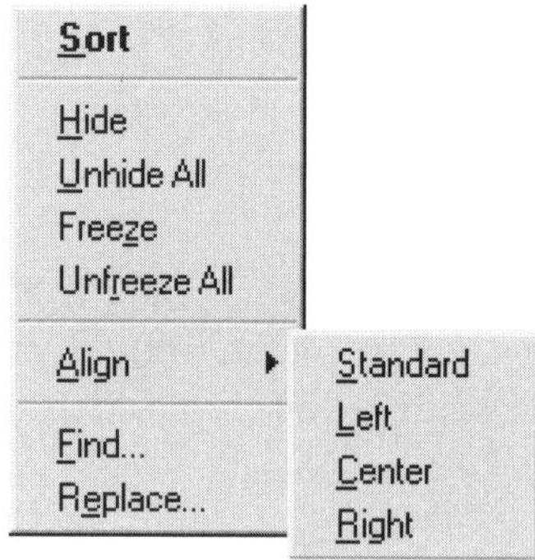

Figure 2.7 *The Column Selector Shortcut Menu*

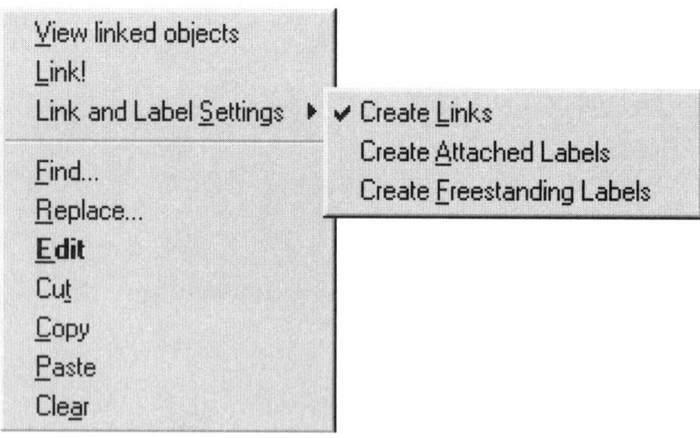

Figure 2.8 *The Cell Shortcut Menu*

DATA MANIPULATION

If you have some experience in using Microsoft Access, you should feel very comfortable in manipulating data in the table grid of the Data View window. The capabilities and behavior of the Data View window are very similar to the Datasheet view in Access. You can

- Modify data directly
- Find and replace data
- Append new rows to the table
- Delete rows from the table

Modifying Data Directly

A table can be opened in the Data View window in either *Edit* mode or *View* mode. In Edit mode, changes can be made to the data directly in the Data View window, provided that you have the appropriate rights and the database provider supports updatable cursors. (A *cursor* is a program's copy of the results of a database command or query.) In View mode, the table is opened in read-only mode and changes cannot be made. When the Data View window is opened in Edit mode, the background color is white. When it is opened in View mode, the background color is gray.

When an attempt is made to open a table in Edit mode, AutoCAD may be able to immediately determine that the table is not truly editable for one reason or another. In this case, it is opened in View mode, and no warning is issued. If AutoCAD cannot determine whether or not the table can truly be edited, it opens the table in Edit mode anyway.

Even when the Data View window is opened in Edit mode, some columns may not be editable for one reason or another. For example, Memo fields and Autonumber fields in Microsoft Access tables cannot be edited directly. If you try to edit a non-editable column, AutoCAD issues a warning as shown in Figure 2.9.

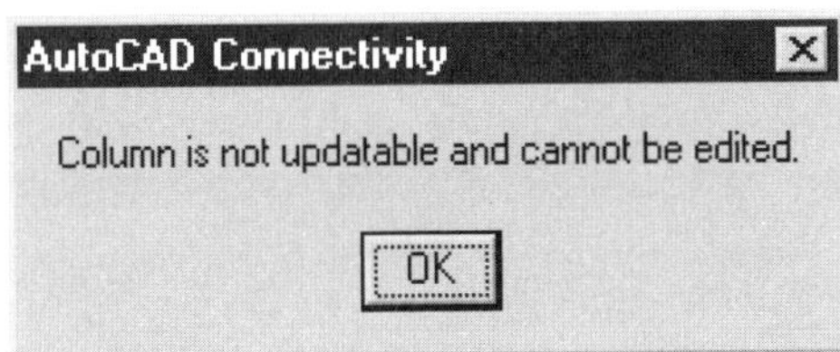

Figure 2.9 *Data View Warning Message*

Changes made to the data in the Data View window are not actually written to the database until you *commit* them. There are three ways to commit changes to the table:

- Close the Data View window
- Open the Data View window with the same table or a different table
- Choose **Commit** from the table selector shortcut menu (this also closes the Data View window)

If you decide that you do not want to commit the changes you have made, you can select **Restore** from the shortcut menu. This will close the Data View window without saving any of your edits.

Using Find and Replace

The Data View window also allows you to make global changes to data using a "find and replace" method. Display the **Replace** dialog box either by choosing **Replace** from the **Data View** menu or by right-clicking on a column and choosing **Replace** from the shortcut menu.

Figure 2.10 shows an example of using the **Replace** feature to change occurrences of "Big Pine Rd" to "Small Pine Rd."

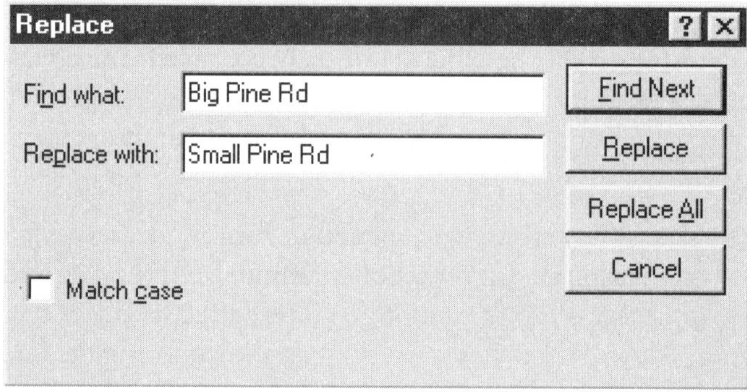

Figure 2.10 *The Replace Dialog Box*

Note: The Find and Replace features of the Data View window operate only on one column at a time. There is no "global" find and replace capability.

Appending New Rows

Appending new rows to a table in the Data View window works pretty much the same as modifying data in existing rows. At the bottom of the table you will find a blank row with an asterisk in the row selector. To add a new row to the table, simply enter the new row's data in the blank row. Once the data have been entered, you will see a "delta" symbol in the row selector. This indicates that the row has been added to the Data View window but has not been written to the database. Committing new rows is accomplished in exactly the same manner as committing modified rows, as described in the previous section.

Note: You can link new rows to objects as they are appended to the table. If you do this, however, remember that if you choose not to commit the rows to the database, the links will remain on the objects but the rows will not exist in the table.

Deleting Rows

To delete a row or multiple rows simply select the rows by clicking in the row selector and press DELETE. You are prompted to confirm the delete operation. The rows disappear from the Data View window, but they are not actually deleted from the database until you close the Data View window or issue a **Commit**.

The following tutorial walks you through the process of modifying data in the Data View window.

TUTORIAL

TUTORIAL 2.1 – MODIFYING DATA IN THE DATA VIEW WINDOW

1. Open parcels3.dwg. This drawing has all the links and queries that you created in Chapter 1.
2. Display the dbConnect Manager, and connect to the parcels data source.
3. Double-click the **PARCEL_DATA** table in the dbConnect Manager window to display the Data View window for this table.
4. Click in the first field under the **ADDRESS** column and type "200." As you are typing, notice that there is a small "pencil" icon in the row selector (on the far left side of the table grid).
5. Press the DOWN ARROW key to advance to the next row. Notice that the row selector now shows a Greek "delta" symbol, indicating that changes have been made to that row. There is also a "delta" symbol in the table selector (in the top left corner of the table grid) indicating that changes have been made to the table, but they have not yet been committed to the database.
6. Continue changing the **ADDRESS** value in the next three rows to "202," "204," and "206" respectively. Your Data View window should now look like Figure 2.11.
7. Right-click in the table selector (in the top left corner of the table grid) to display the table selector shortcut menu (shown earlier in Figure 2.5). Using this menu gives you the option to commit changes or restore the table to its original state. Either option will close the Data View window.
8. Choose **Commit** to commit the changes and close the Data View window.

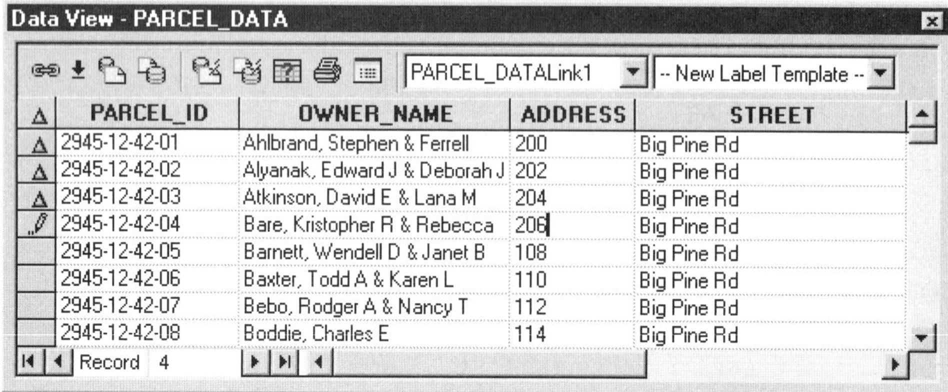

Figure 2.11 *Modifying Data in the Data View Window*

CHANGING THE APPEARANCE OF THE DATA VIEW WINDOW

There are several ways you can modify the display characteristics of the Data View window, including text justification, sort order, column order, and font. These changes apply only to the current "session" of the Data View window. Once the Data View window is reopened, the changes are lost, and the window takes on its default appearance.

Sorting Data

To sort the data on a single column, you can simply double-click the column header. Double-clicking again will reverse the sort order. If you want to sort on more than one column, you must use the **Sort** dialog box. Display the **Sort** dialog box by choosing **Sort** from the column selector shortcut menu. Figure 2.12 shows how the **Sort** dialog box could be used to sort the **PARCEL_DATA** table by **STREET** and ascending order, and then by **ACRES** in descending order.

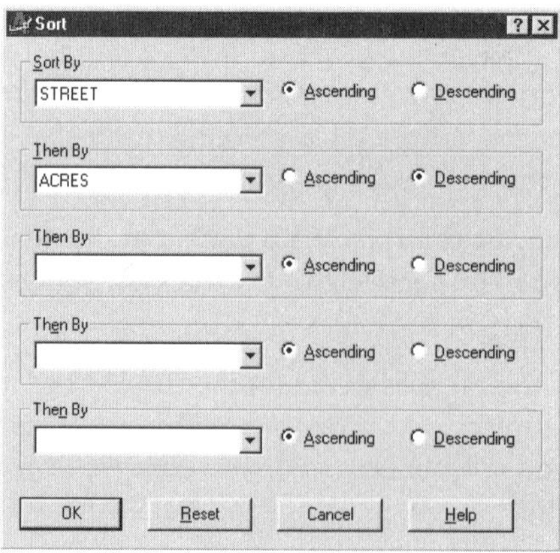

Figure 2.12 *The Sort Dialog Box*

Hiding and Freezing Columns

Hide and **Freeze** options are also on the column selector shortcut menu. The **Hide** option temporarily hides a column from view. The **Freeze** option takes selected columns, places them at the left side of the data view window, and "freezes" them. The frozen columns stay visible even when you scroll horizontally across other columns. For example, if you always wanted to see the **OWNER_NAME** column while scrolling other columns into view, you would select the **OWNER_NAME** column and choose **Freeze** from the shortcut menu.

Changing the Text Alignment

The text alignment (or justification) can be controlled through the column selector shortcut menu (Figure 2.7). This menu item has a submenu with the following four options:

> **Standard**—This is the default. Character data is left justified, and numeric data is right justified.
>
> **Left**—The column is left justified.
>
> **Center**—The column is center justified.
>
> **Right**—The column is right justified.

Adjusting Column Widths

Adjust the widths of columns in the Data View window simply by selecting the line between two column selectors. Figure 2.13 illustrates this.

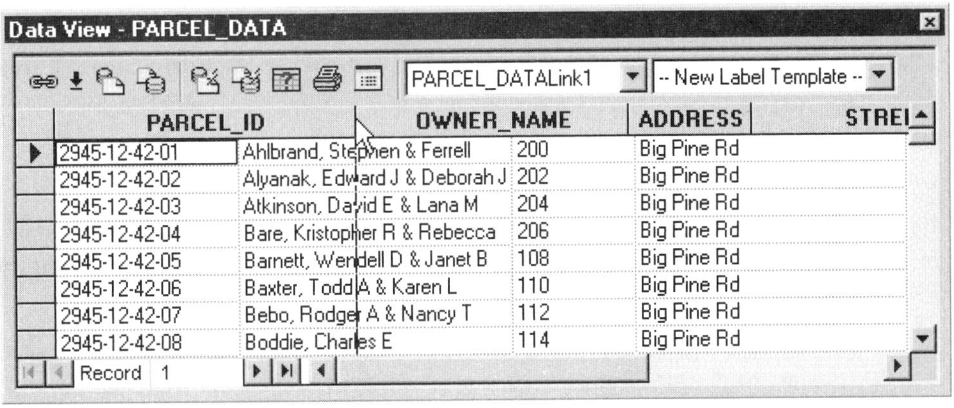

Figure 2.13 *Adjusting Column Widths*

Changing the Font

The characteristics of the font used in the Data View window can be modified through the **Format** dialog box. Display the **Format** dialog box by choosing **Format** from either the table selector shortcut menu or the **Data View** menu.

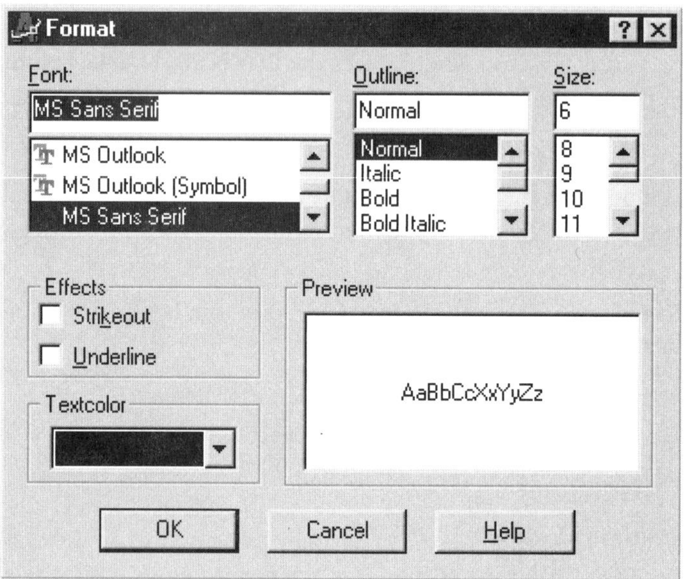

Figure 2.14 *The Data View Format Dialog Box*

Note: With the exception of font color, changes to the font characteristics are only active for the current session of the Data View window. Font color changes stay active for the entire AutoCAD session.

VIEWING LINKED RECORDS AND LINKED OBJECTS

One of the most powerful features of the Data View window is the ability to instantly visualize the connectivity between the table and objects. This feature gives you the following capabilities:

- Highlight objects in the drawing by selecting the linked rows. There are several ways to do this. First, select one or more rows in the Data View window, and then do one of the following:
    - Choose the **View Linked Objects in Drawing** icon on the Data View toolbar.
    - Choose **View linked objects** from the row selector shortcut menu.

    - Choose **View Linked Objects** from the **Data View** menu.
    - Double-click a row selector.
- Highlight records in the Data View by selecting the linked objects. To do this, first select one or more linked objects, and then choose the **View Linked Records in Data View** icon on the Data View toolbar, or choose **View Linked Records** from the **Data View** menu.

You can also configure the Data View window to *automatically* view either linked rows or linked objects when the corresponding objects or records are selected. To activate either of these automatic features, choose **AutoView Linked Objects** or **AutoView Linked Records**. Both these options are available on the Data View toolbar and on the **Data View** menu.

CONTROLLING DATA VIEW OPTIONS

Several options are available that control the behavior of the Data View window when you view linked records and linked objects. These options are controlled through the **Data View and Query Options** dialog box, shown in Figure 2.15.

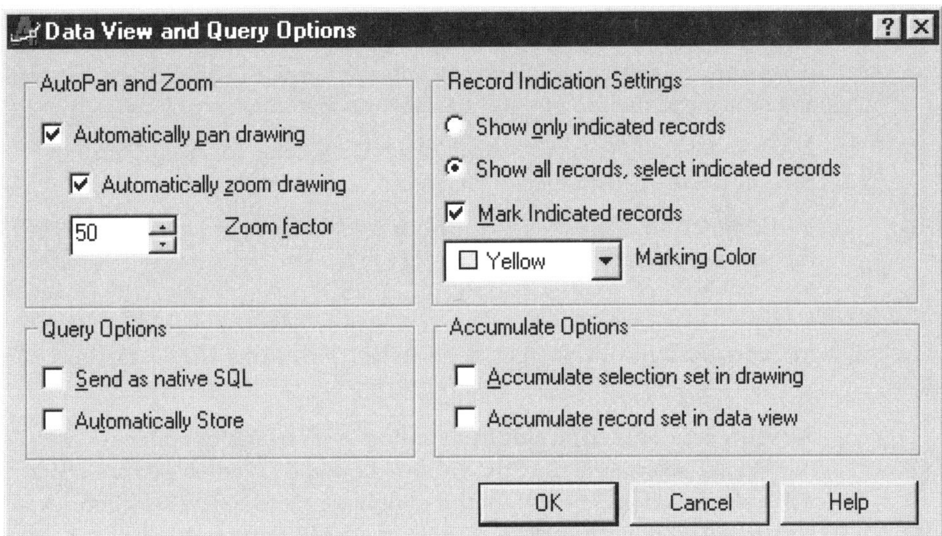

Figure 2.15 *The Data View and Query Options Dialog Box*

The **Data View and Query Options** dialog box contains the following options related to the Data View window:

AutoPan and Zoom

This area allows you to control the behavior of the **View Linked Objects** feature. You have the following options:

Automatically Pan Drawing—Pans the AutoCAD drawing to bring the selected objects to the center of the screen.

Automatically Zoom Drawing—This option is only available if the **Pan** option is also selected. When this is active, AutoCAD zooms the drawing so that all the selected objects are visible.

Zoom Factor—This represents the percentage of the drawing area that the selected objects will occupy. You can choose a value between 20 and 90 percent. The default is 50 percent.

Record Indication Settings

This area controls how records in the Data View window are highlighted when the **View Linked Records** feature is activated. You have the following options:

Show Only Indicated Records—Displays *only* the records that are linked to the selected AutoCAD objects.

Show All Records, Select Indicated Records—Displays *all* records in the table. Records that are linked to the selected AutoCAD objects are selected and brought to the top of the Data View window.

Mark Indicated Records—Applies a marking color to linked Data View records to clearly differentiate them from records without links.

Marking Color—Specifies the marking color to apply to linked Data View records. The default color is yellow.

Accumulate Options

The following options control how selection sets are built using the Data View window:

Accumulate Selection Set in Drawing—When this option is set, as records are selected in the Data View window, the linked objects are added to the selection set in AutoCAD. If this option is cleared, a new selection set is created each time you select a new set of Data View records.

Accumulate Record Set in Data View—When this option is set, as objects are selected in AutoCAD, additional records are selected in the Data View window. This option is only available when **Show Only Indicated Records** is selected.

TUTORIAL 2.2 – VIEWING LINKED RECORDS AND LINKED OBJECTS

1. If you did not continue from the previous tutorial, open parcels3.dwg, display the dbConnect Manager, and connect to the parcels data source.
2. Double-click the **PARCEL_DATA** table in the dbConnect Manager window to display the Data View window for this table.
3. Dock the Data View window at the bottom of the AutoCAD window.
4. From the **Data View** menu, choose **Options**....
5. Under **AutoPan and Zoom**, make sure that **Automatically Pan Drawing** and **Automatically Zoom Drawing** are both selected.
6. Set the zoom factor to 60 percent.
7. Under **Record Indication Settings**, choose **Show only indicated records**.
8. Choose **OK**.
9. In the Data View window, double-click a row selector, and notice that AutoCAD zooms in on the linked parcel.
10. Select multiple rows by holding down SHIFT.
11. Right-click the row selector, and choose **View linked objects** from the menu. AutoCAD zooms the selected objects into view.
12. From the **View** menu, choose **Zoom** and then **Extents**.
13. From the **Data View** menu, choose **AutoView Linked Records**.
14. Select one or more parcel polylines in the AutoCAD window. Note that only the linked records are displayed in the Data View window.
15. To reset the Data View window to show all records, simply double-click the **PARCEL_DATA** table in the dbConnect Manager window.

PRINTING THE DATA VIEW

At any time, the contents of the Data View window can be sent to your printer. The printing capability of the Data View window is not intended to be a "reporting" capability. You have very little control over the appearance of the output. It is simply a way to make a hard copy of the current view in the Data View window.

Many of the display characteristics of the Data View window are carried over to the printed output, including

- Row sort order
- Column order

- Column hide/show status (hidden columns are not printed)
- Column widths
- Font style and size
- Text alignment

You can preview the printed output by choosing **Print Preview** from either the **Data View** menu or the table selector shortcut menu. The capabilities and behavior of the print preview are very similar to that of other Windows applications.

To send the Data View contents to the printer, you can use the **Print Data View** icon on the Data View toolbar. You can also choose **Print** from either the **Data View** menu or the table selector shortcut menu.

USING THE CLIPBOARD

The Data View grid provides you with the standard clipboard options of *cut*, *copy*, and *paste*. This could be useful, for example, if you wanted to take a group of cells and bring them into a spreadsheet program such as Microsoft® Excel.

CREATING COMPLEX QUERIES

In the first chapter, we touched on the basics of the Query Editor. As you recall, there are four different ways to create queries:

Quick Query—For very simple queries

Range Query—To query based on a range of values

Query Builder—For more complex queries

SQL Query—For advanced users who just want to type the necessary SQL statements to create the query

In Chapter 1, we showed examples of using the first two methods. Here, we will focus on the second two.

THE QUERY BUILDER

The Query Builder provides an easy-to-use interface for creating fairly complex queries. You can use multiple columns and expressions to formulate your query, and you can group expressions together using parentheses.

The following tutorial illustrates the capabilities of the Query Builder.

TUTORIAL 2.3 – USING THE QUERY BUILDER

1. If you did not continue from the previous tutorial, open parcels3.dwg, display the dbConnect Manager, and connect to the parcels data source.
2. Right-click the **PARCEL_DATA** table, and select **New Query** from the shortcut menu.

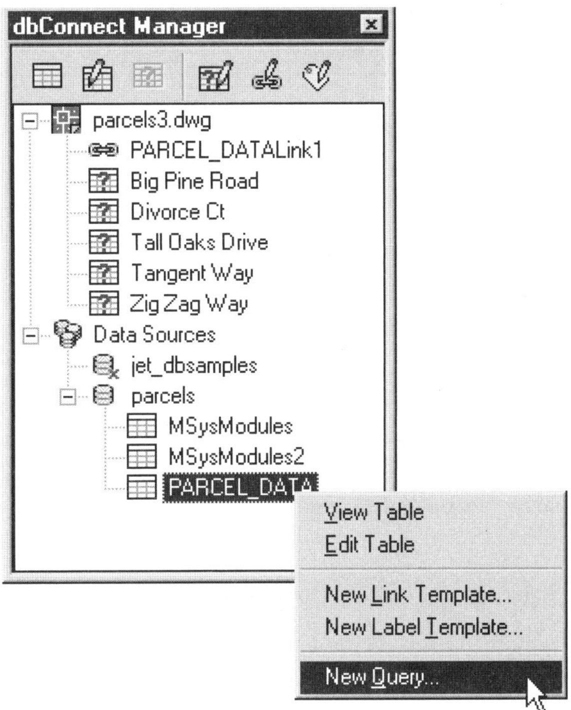

Figure 2.16 *The Table Shortcut Menu*

3. In the **New Query** dialog box, leave the default query name as **PARCEL_DATAQuery1** and click **Continue**.

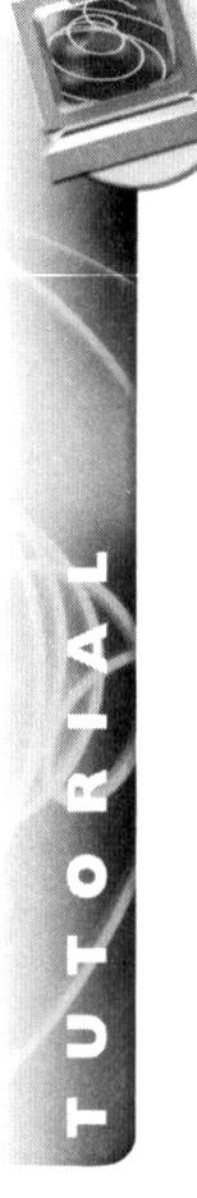

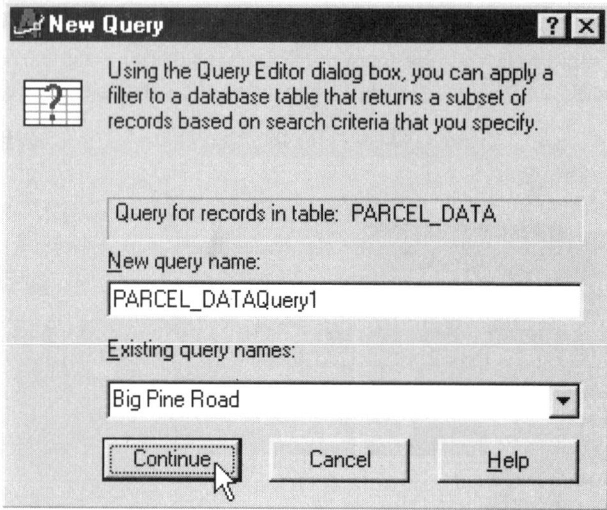

Figure 2.17 *The New Query Dialog Box*

4. In the **Query Editor** dialog box, select the **Query Builder** tab. When this tab is activated, the Query Editor shows a grid area with the headings **Field**, **Operator**, **Value** and **Logical**, as shown in Figure 2.18.

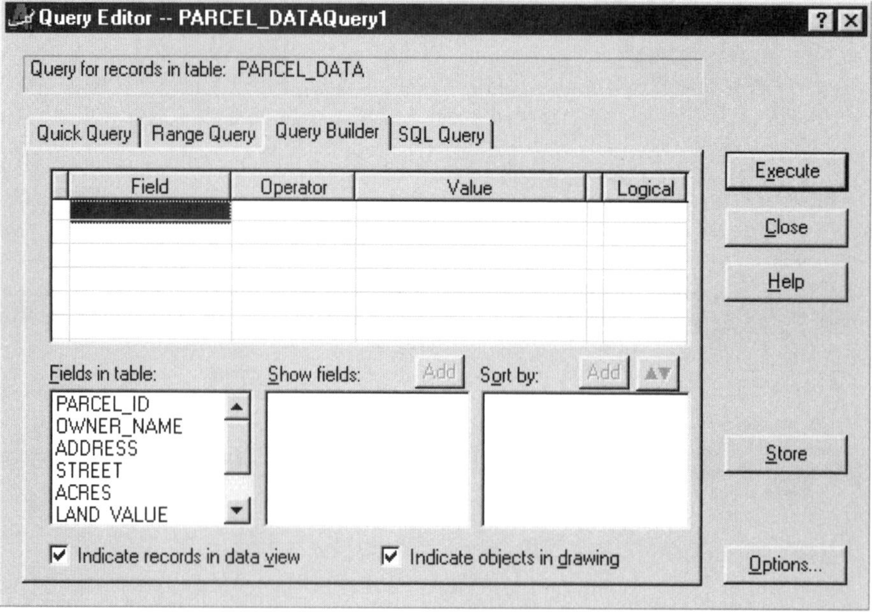

Figure 2.18 *The Query Builder*

5. In the first row of the grid, click in the first cell under the **Field** column. A pick list will appear with a list of available fields.
6. Choose **STREET** from the list of fields, as shown in Figure 2.19.

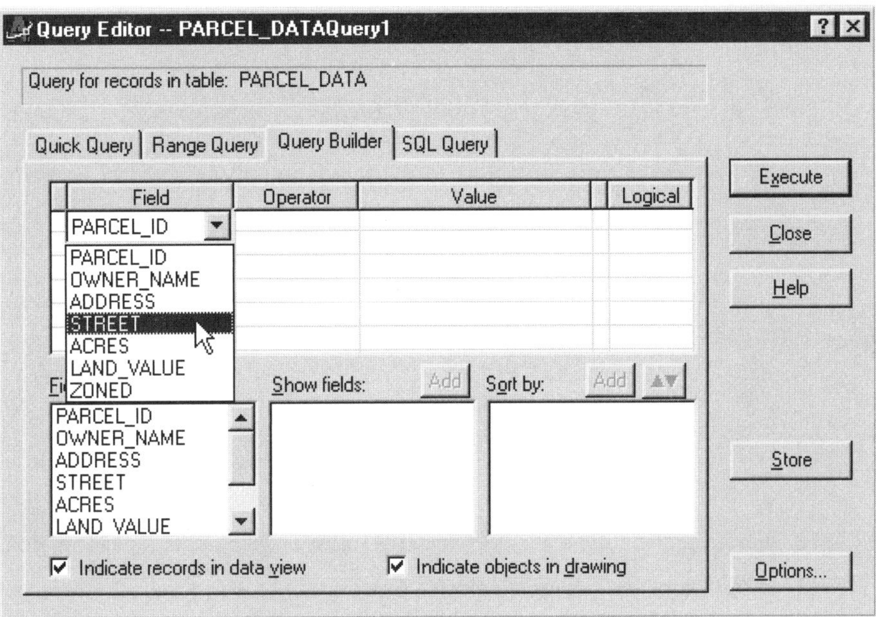
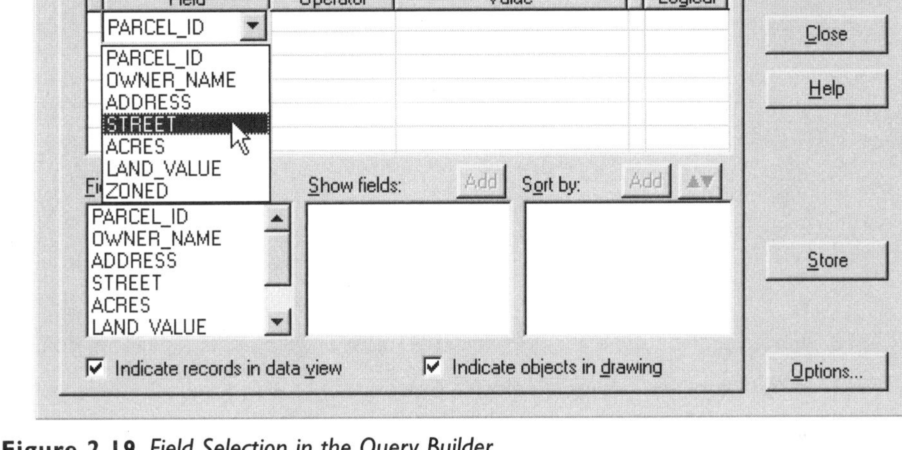

Figure 2.19 *Field Selection in the Query Builder*

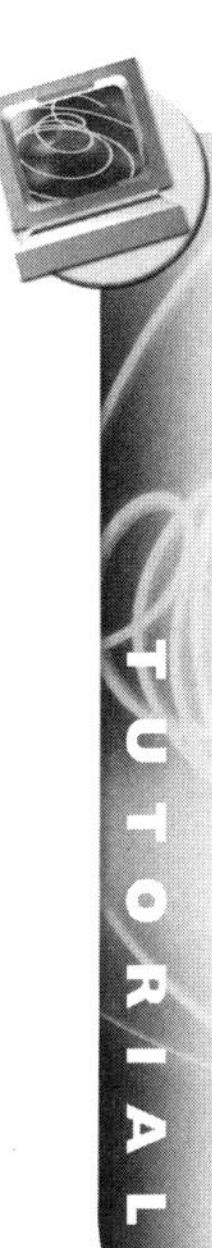

7. In the **Operator** column, choose **= Equal**.
8. Click in the **Value** column, and an edit box will appear with an ELLIPSIS ("**...**") button.
9. Click the ELLIPSIS button and a dialog box will appear (Figure 2.20) with the list of available values for the **STREET** field.

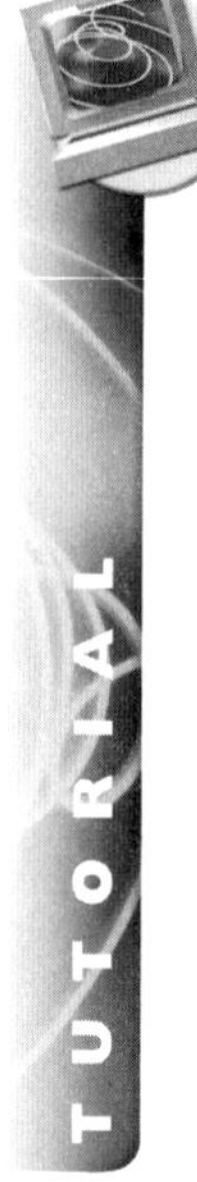

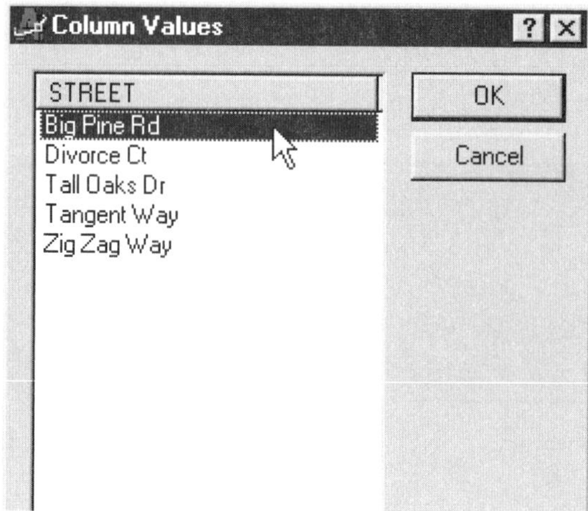

Figure 2.20 *Column Values Dialog Box*

10. Choose **Big Pine Rd** from the list and click **OK**.
11. Click in the **Logical** column until you see the word "Or."
12. In the second row in the grid, select **STREET** again under the **Field** column.
13. In the **Operator** column, select **= Equal**.
14. In the **Value** column, use the ellipsis button and select **Divorce Ct** from the list and click **OK**.

Note: Notice that there are two narrow, unnamed columns in the grid: one just to the left of the **Field** column, and one just to the right of the **Value** column. These can be used to parenthetically group lines together. Clicking in these cells will toggle the parentheses on and off.

15. Group these first two lines together by clicking in the parentheses cell in the first row, just to the left of the **Field** column, and again in the second row, just to the right of the **Value** column (Figure 2.21).
16. Click twice in the **Logical** column, so that the word "And" appears.
17. In the third line of the grid, select **LAND_VALUE** under the **Field** column.
18. In the **Operator** column, select **< Less than**, as shown in Figure 2.22.
19. Type "25000" in the **Value** column.
20. In the **Fields in Table** list, choose **PARCEL_ID** and click the **Add** button located just above the empty list box labeled **Show fields**. This adds **PARCEL_ID** to the list of displayed fields.

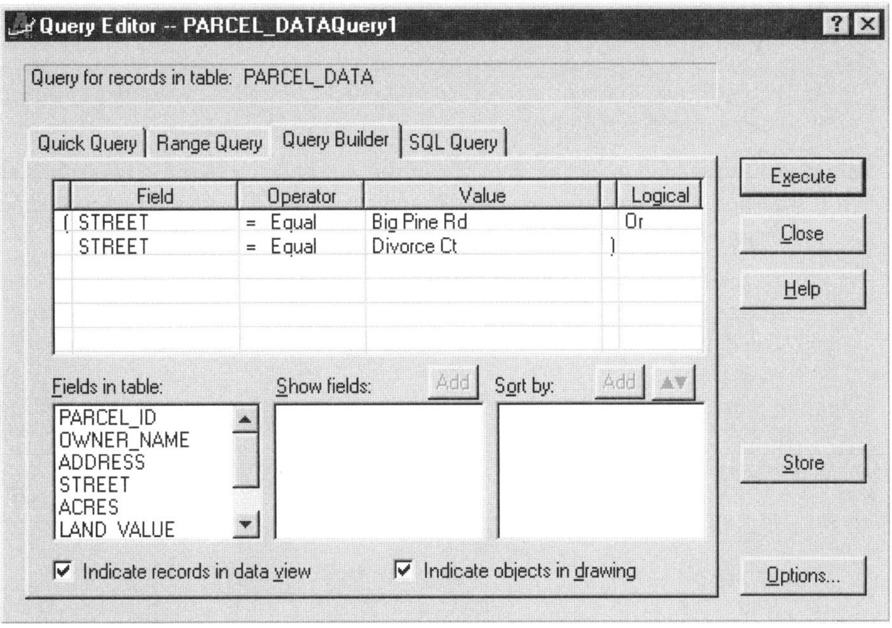

Figure 2.21 *Parenthetically Grouped Expressions*

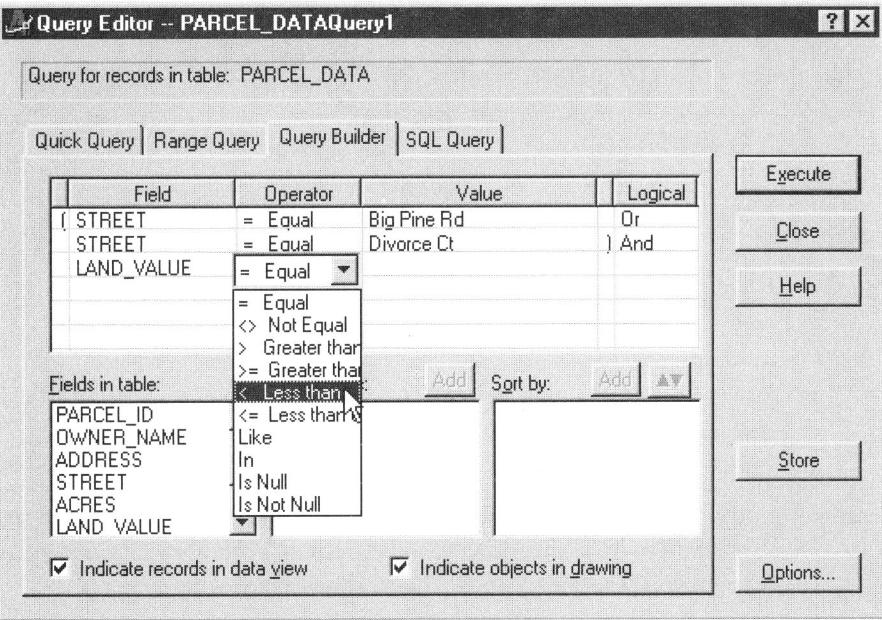

Figure 2.22 *Choosing an Operator*

21. Continue clicking the **Add** button until all the fields have been added to the list.
22. In the **Show fields** list, select **LAND_VALUE** and click the **Add** button next to the **Sort by** list.
23. Click the button just to the right of the **Add** button that has two triangles pointed in opposite directions. This changes the sort order from ascending to descending.
24. At the bottom of the dialog box, make sure that both the **Indicate records in data view** and **Indicate objects in drawing** check boxes are selected. The **Query Editor** dialog box should look like Figure 2.23.

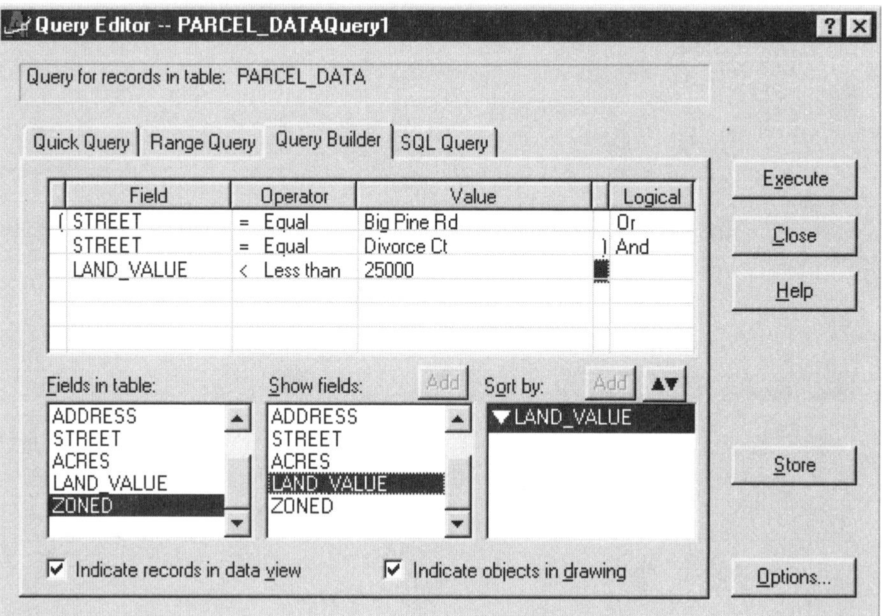

Figure 2.23 *Using the Query Builder*

25. Click **Execute.** The Data View window is displayed. Keep the Data View window open for the next tutorial.

When the query is issued, the resulting rows are shown in the Data View window, and the corresponding linked objects are highlighted in the drawing. If you want to return directly to the **Query Editor** dialog box to refine or modify your query, you can choose the **Return to Query** icon on the **Data View** window toolbar.

SQL QUERY

SQL Query, the fourth tab on the **Query Editor** dialog box, provides the most flexible way to issue queries on your data from within AutoCAD.

This tab consists primarily of a text area in which any valid SQL statement can be entered. This is the only tab in the **Query Editor** dialog box in which a query can be formulated that contains more than one table. It is also the only tab that allows a non–row-returning SQL statement, such as **INSERT**, **UPDATE**, or **DELETE**.

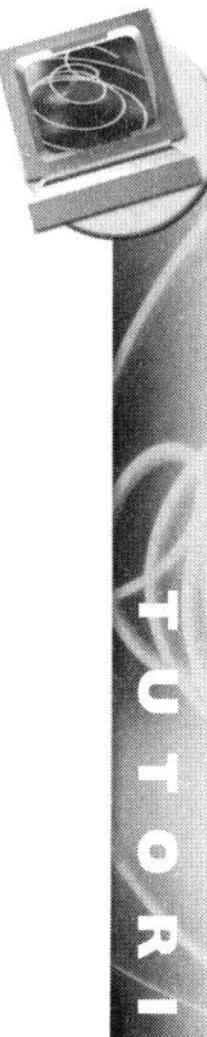

Many of the same helpful query-building controls that were found on the other tabs are still available and can assist in building a query. For example, you can select items from lists of tables or fields, and the selected items will be appropriately placed within the query.

The following tutorial shows how this tab is used.

TUTORIAL 2.4 – USING THE SQL QUERY TAB

1. On the Data View window that was displayed in the previous tutorial, choose the **Return to Query** icon in the toolbar. The **Query Editor** dialog box is displayed with the current query shown.
2. Choose the **SQL Query** tab. The query that we just executed is shown using SQL.

The SQL query shown in Figure 2.24 should read as follows:

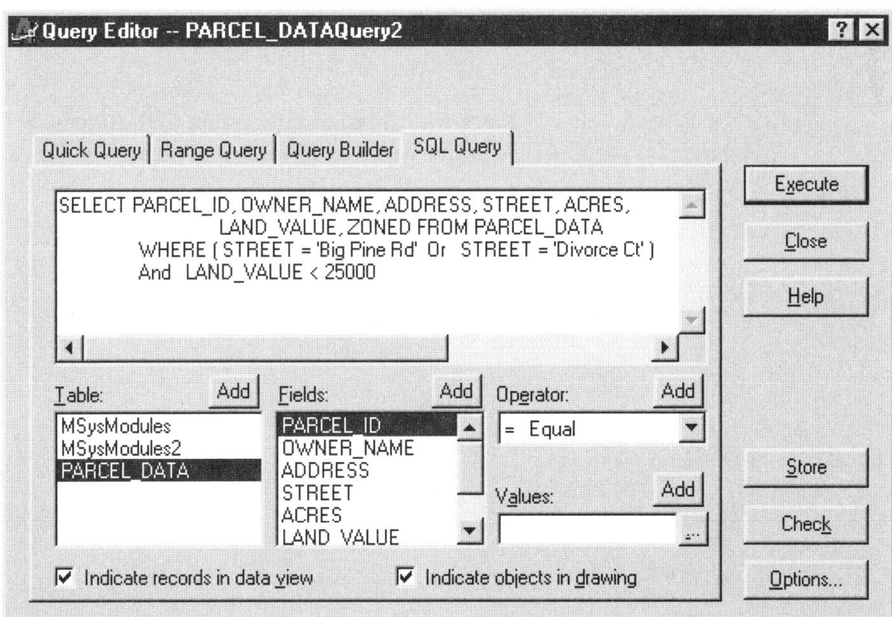

Figure 2.24 *SQL Query*

```
SELECT PARCEL_ID, OWNER_NAME, ADDRESS, STREET,
ACRES, LAND_VALUE, ZONED FROM PARCEL_DATA
   WHERE ( STREET = 'Big Pine Rd'  Or  STREET = 'Divorce Ct' )
And  LAND_VALUE < 25000
                  ORDER BY LAND_VALUE DESC
```

3. Change the **LAND_VALUE < 25000** to **LAND_VALUE > 25000.** You may need to scroll the query window horizontally to see this part of the query. You can also resize the **Query Editor** dialog box.
4. Choose the **Query Builder** tab. A warning message is displayed (Figure 2.25) to inform you that choosing a "previous" tab (meaning any tab to the left of the current tab) will reset the **Query Editor** dialog box to its default state.
5. Choose **No** on the **Warning Message** dialog box to return to the Query Editor.

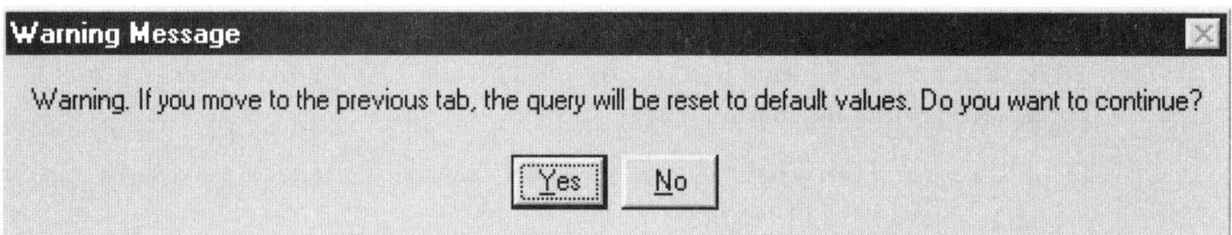

Figure 2.25 *Warning Message*

6. Choose **Check.** This will check the syntax of the current SQL statement, without actually executing it. If the syntax is correct, the message in Figure 2.26 is displayed.

Figure 2.26 *SQL Syntax Message*

7. Choose **OK** to clear the SQL syntax message.
8. Choose **Execute**.

In order to get the most benefit from the SQL Query tab, you should first have a good understanding of SQL. In Chapter 4, we will focus exclusively on SQL and further demonstrate this tab of the **Query Editor** dialog box.

QUERY PROCESSING OPTIONS

There are a couple of options for SQL query processing that can be controlled from the **Data View and Query Options** dialog box (show earlier in Figure 2.15). Figure 2.27 shows the portion of this dialog box that contains the options for query processing.

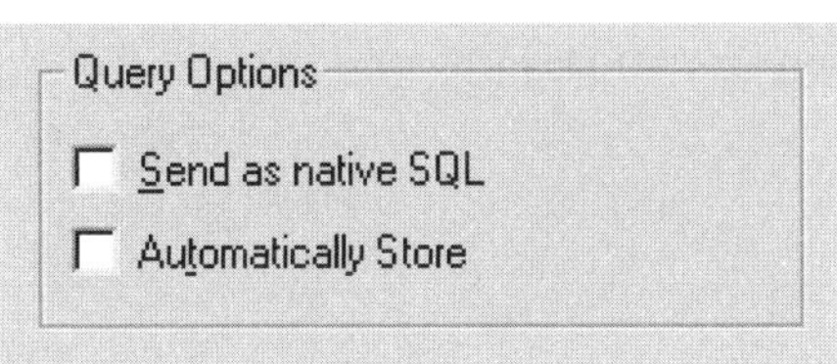

Figure 2.27 *Query Options*

The options are described as follows:

> **Send as Native SQL**—Queries are sent directly to the data provider without any preliminary syntax checking. You can use this option to issue non-standard SQL commands to a database.
>
> **Automatically Store**—Queries are automatically stored in the current drawing at the time they are executed.

SHARING QUERIES WITH OTHER DRAWINGS

At times, you may have a set of queries in a drawing that you want make available to other drawings or to other users. DbConnect provides a query import/export capability for this purpose.

Import/Export

Issue the import and export commands by right-clicking a drawing node in the dbConnect Manager window and choosing **Import Query Set** or **Export Query Set**. These commands are also available from the **Queries** submenu of the **dbConnect** menu.

Exporting a query set stores all queries in the current drawing to a file with a *.dbq* extension. This file can then be imported to another drawing. If you import a query set that contains queries that have the same name as queries already defined in the current drawing, you will be prompted to enter a new name for each duplicate query, as shown in Figure 2.28.

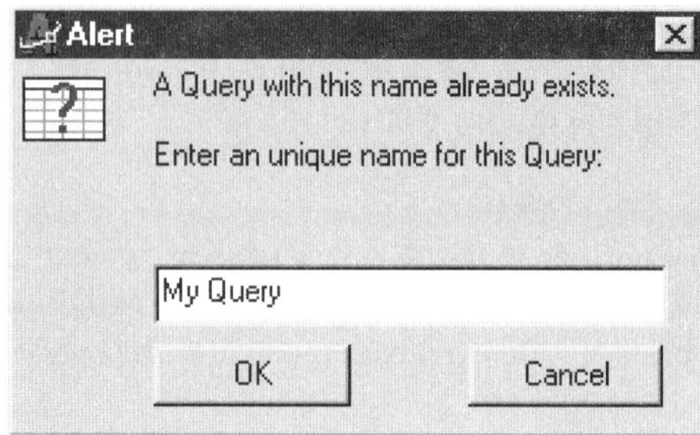

Figure 2.28 *Import Query Set Warning Message*

Drag and Drop

If you have multiple drawings open in AutoCAD, and you want to copy a query from one drawing to another, you can simply drag the query from one drawing node to another in the dbConnect Manager window.

SELECTING OBJECTS USING LINK SELECT

The Link Select feature of dbConnect is very similar to the Query feature. Access it from the dbConnect menu by selecting **Links** and then **Link Select** (Figure 2.29).

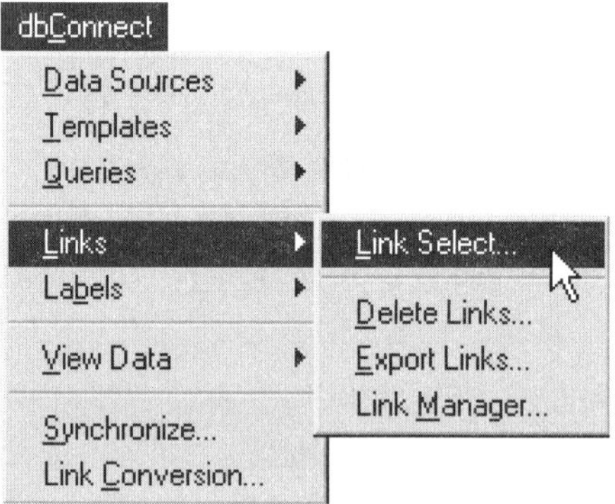

Figure 2.29 *Link Select on the dbConnect Menu*

Link Select has the following features not found in the Query Editor:

Iterative Querying—Link Select allows you to continuously refine your query by combining the results of one query with the results of another. The combined query can then be combined with another query, and so on. Queries can be combined through the standard Boolean operations of union, intersection, and subtraction.

Graphical Selection—Link Select allows you to use both database queries and manual object selection.

Multiple Link Templates—Link Select allows you to combine results of queries using more than one link template.

However, Link Select is missing the following features of the Query Editor:

Store—Since Link Select uses an iterative query process, it does not allow you to store queries in the drawing.

SQL Query—Link select does not provide a free-form query mechanism. This is because tables queried with Link Select must be associated with a link template, and only "select" operations are permitted.

Each interface has its advantages in certain situations. For example, if you find you are using the Query Editor to build selection sets in AutoCAD, and you don't need to store them for later use, you should probably be using Link Select.

The **Link Select** dialog box, shown in Figure 2.30, looks very similar to the **Query Editor** dialog box. The **Quick Query**, **Range Query**, and **Query Builder** tabs are exactly the same.

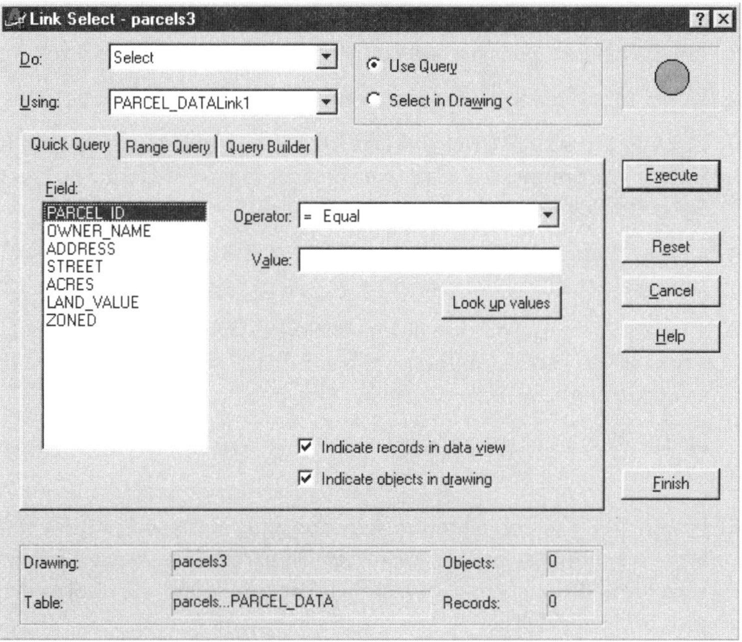

Figure 2.30 *The Link Select Dialog Box*

Link Select uses the concept of a *result set*. Each time a query is executed, it creates a result set—a series of database rows and/or AutoCAD objects that satisfy the query. Once a result set has been created, it becomes known to the Link Select dialog box as set A. You can then issue an additional query, known as **B**, and combine the two queries using the following Boolean operations:

Union—*Result = A or B*

Intersect—*Result = A and B*

Subtract A-B—*Result = A and not B*

Subtract B-A—*Result = B and not A*

The result then becomes the new A, which can then be combined with another query **B**, and so on.

To illustrate this, let's look at an example. Suppose our parcels drawing represents a small town, and you want to know which parcels on the east side of town are smaller than a quarter of an acre. This query requires the use of the **Link Select** dialog box, because it includes a graphical query. There is no information in the database that tells us that a particular parcel is on the "east" side of town.

The following tutorial takes you through this example.

TUTORIAL 2.5 – USING LINK SELECT

1. If you did not continue from the previous tutorial, open parcels3.dwg, display the dbConnect Manager, and connect to the parcels data source.
2. Zoom to the extents of the drawing.
3. Right-click the **PARCEL_DATALink1** link template, and choose **Link Select...** from the link template shortcut menu (Figure 2.31). The **Link Select** dialog box is displayed, as shown earlier in Figure 2.30.
4. Choose **Select in Drawing <** and click the **Select** button. This will hide the **Link Select** dialog box and prompt you to select objects in the drawing.
5. Imagine a vertical line that divides the parcel map into two equal halves, and select all the parcels on the right hand side of that line using the window selection method.

Note: Objects that are not linked with the current link template will automatically be filtered out of the selection.

6. Press ENTER to complete the selection and return to the **Link Select** dialog box. This should select approximately 34 parcels. The exact number of objects selected and the number of matching records found are displayed at the bottom of the dialog box.

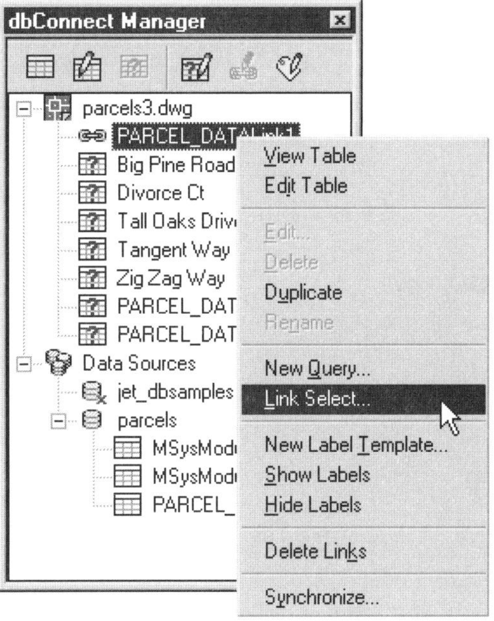

Figure 2.31 *Link Select on the Link Template Shortcut Menu*

7. Choose **Intersect** from the **Do** pick list at the top of the dialog box.
8. Choose **Use Query**, and make sure the **Quick Query** tab is active.
9. In the **Field** list, select **ACRES**.
10. In the **Operator** pick list, choose **< Less Than**.
11. Type ".25" in the **Value** field. The **Link Select** dialog box should look like Figure 2.32.
12. Click **Execute**. Notice the change in the number of objects and number of records displayed at the bottom of the dialog box.
13. Make sure that the **Indicate records in data view** and **Indicate objects in drawing** check boxes are both selected.
14. Click **Finish**. The objects that met the criteria are highlighted, and the Data View window appears with the corresponding rows displayed.

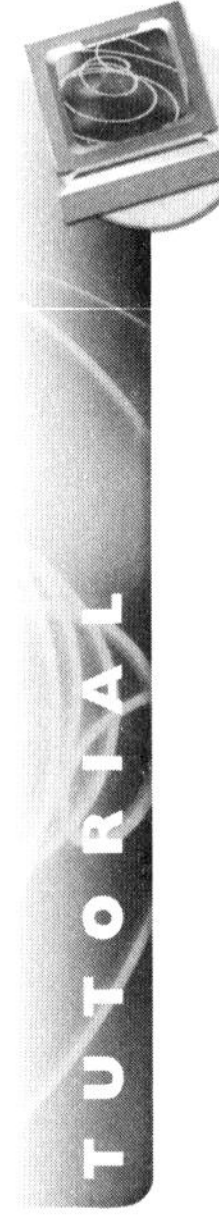

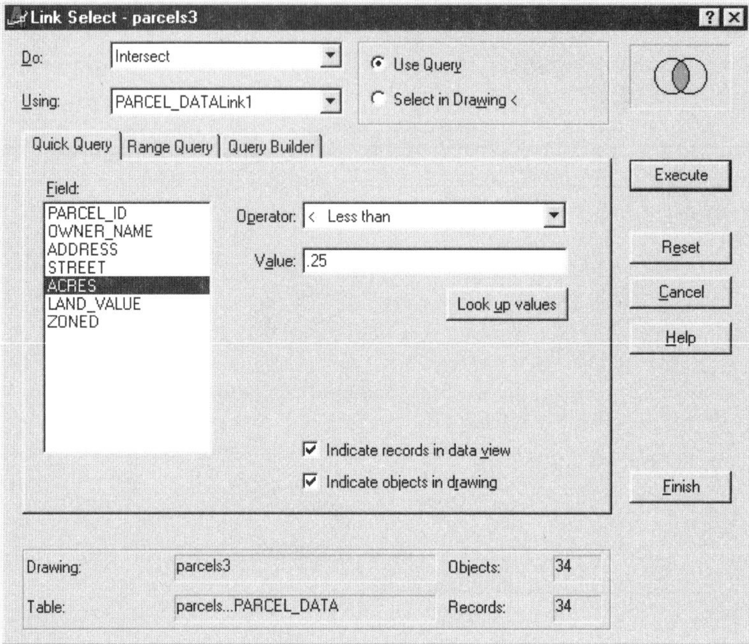

Figure 2.32 *The Link Select Dialog Box*

EXPORTING LINKS

This feature of dbConnect gives you the ability to export link information from a selection set of objects. There are three methods of export available:

- A comma-delimited file
- A space-delimited file
- A new table in your database

The first two methods create a file on your system, while the third method actually creates a new table in your database. When you select objects for the **Export Links** command, you can choose to export one or more columns from the linked table. In addition to the columns you select, AutoCAD adds a column for the entity handle.

This feature is especially useful if you want to store information about a selection set of objects for later use. For example, suppose you found that you were frequently issuing queries on the parcels on the east side of the town. It would be much easier if this information were stored in the database, eliminating the need to select the parcels every time you wanted to create a query.

With the export feature, you can select the objects once and create a table in your database containing the parcel IDs of the selected objects. The new table could then be used in other database queries that need to include those criteria.

The following tutorial explains how this would be accomplished.

TUTORIAL 2.6 – EXPORTING LINKS TO A DATABASE TABLE

1. If you did not continue from the previous tutorial, open parcels3.dwg, display the dbConnect Manager, and connect to the parcels data source.
2. Zoom to the extents of the drawing.
3. Press ESC a few times to be sure that no objects are selected.
4. From the **dbConnect** menu, choose **Links** and then **Export Links** (Figure 2.33).

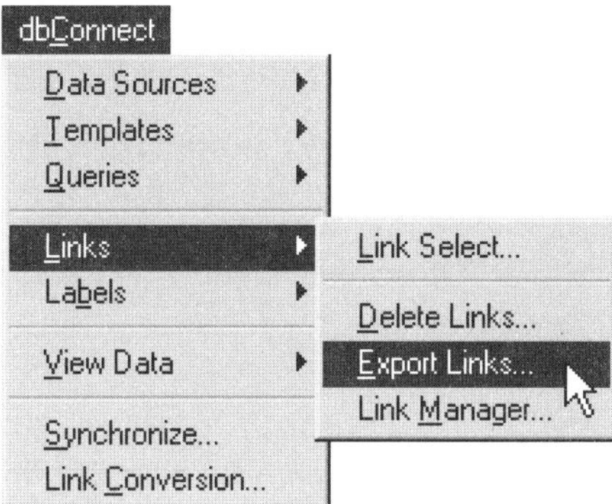

Figure 2.33 *Export Links on the dbConnect Menu*

5. Imagine a vertical line that divides the parcel map into two equal halves, and select all the parcels on the right hand side of that line using the window selection method.
6. Press ENTER to complete the selection. This will display the **Export Links** dialog box, as shown in Figure 2.34.
7. In the **Save as type** pick list, choose **native database format**.
8. Type "**EAST_SIDE** " in the **File name** field.
9. Click **Save.** After a short delay, the dialog box closes, and the **EAST_SIDE** table appears under the parcels data source node in the dbConnect Manager window.

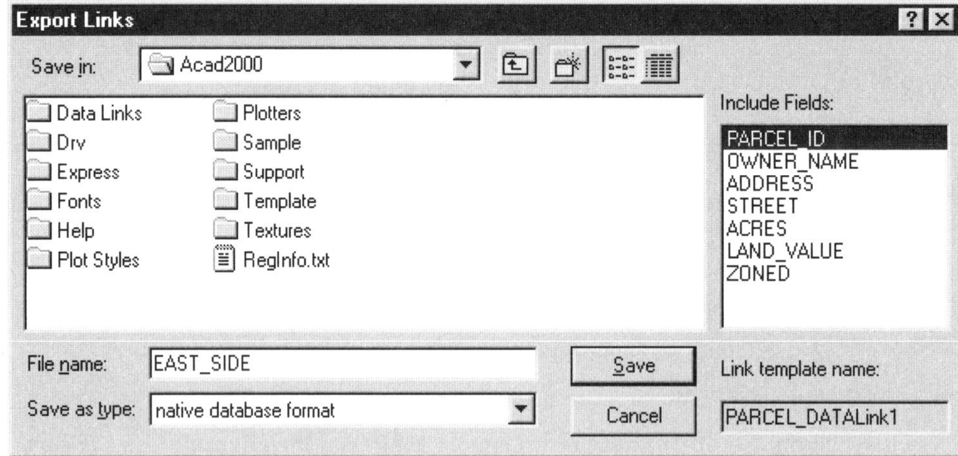

Figure 2.34 *The Export Links Dialog Box*

10. Right-click the **PARCEL_DATA** table and choose **New Query**.
11. Leave **PARCEL_DATAQuery1** as the default name and choose **Continue**.
12. Choose the **SQL Query** tab.
13. Type the following SQL statement in the text area:

```
SELECT * FROM PARCEL_DATA
WHERE ACRES < 0.25
AND PARCEL_ID IN (SELECT PARCEL_ID FROM EAST_SIDE)
```

14. At the bottom of the dialog box, make sure that both the **Indicate records in data view** and **Indicate objects in drawing** check boxes are selected. The **Query Editor** dialog box should look like Figure 2.35.
15. Choose **Execute**.

This should produce the same result as the previous tutorial, which used **Link Select**. This tutorial also demonstrated some of the more advanced features of SQL, which will be described in detail in Chapter 4.

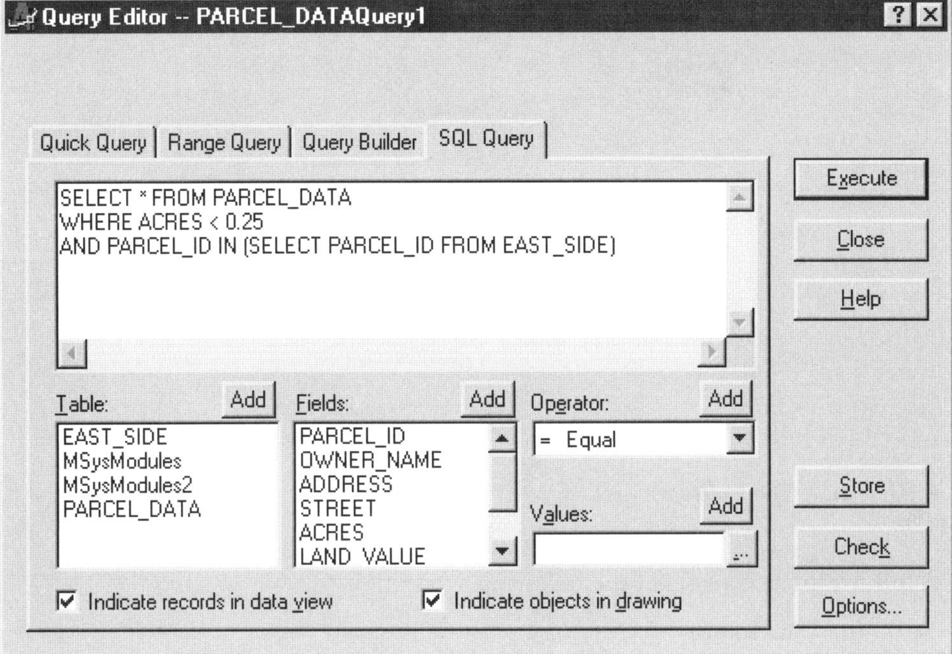

Figure 2.35 *The Query Editor Dialog Box*

MANAGING LINK TEMPLATES

Figure 2.36 shows the **Templates** submenu of the AutoCAD **dbConnect** Menu. This menu gives you access to all the tools you need to manage both link templates and label templates.

MODIFYING A LINK TEMPLATE

Notice in the **Templates** menu (Figure 2.36) that there are two ways to modify a link template. One is called **Edit Link Template**, and the other is called **Link Template Properties**. It is important to understand the distinction between the two methods.

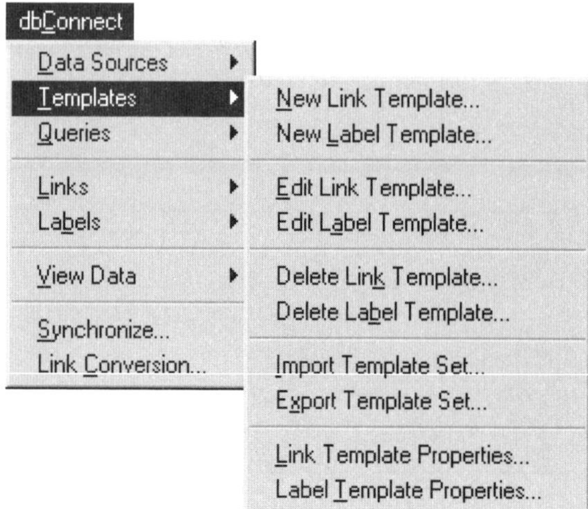

Figure 2.36 *The Templates Menu*

A link template consists of the following properties:

- Data Source
- Catalog
- Schema
- Table
- Key column(s)

To modify the data source, catalog, schema, or table, you use the **Link Template Properties** option.

To modify the key columns, you use the **Edit Link Template** option.

Modifying the key columns has a direct impact on linked objects because they define the number of key values attached to each object, as well as the column names and data types. Therefore, the key columns can only be modified if no links exist for the link template.

On the other hand, the data source, catalog, schema, and table have no direct impact on linked objects. These properties can be changed at any time through the **Link Template Properties** dialog box.

Modifying Link Template Properties

Figure 2.37 shows the **Link Template Properties** dialog box. To display this dialog box, select **Link Template Properties** from the **Templates** submenu of the **dbConnect** menu.

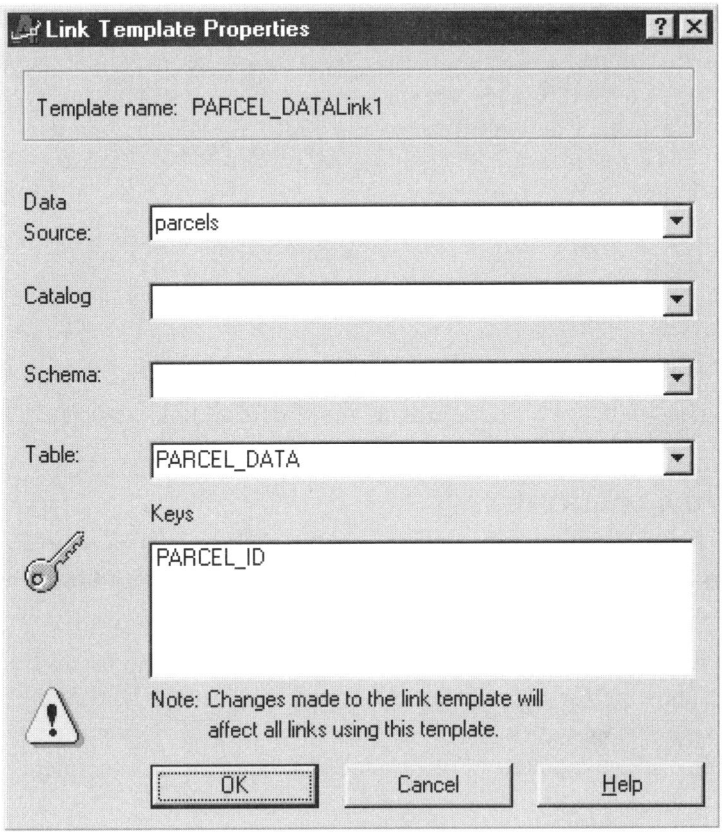

Figure 2.37 *The Link Template Properties Dialog Box*

You may not need to use this dialog box very often, but if you ever do, it can prove to be one of the most useful features of dbConnect. For example, suppose you have been using a Microsoft Access database, and you need to upgrade to SQL Server. This dialog box gives you the flexibility of changing the "path" to the database, without forcing you to re-link your objects.

Modifying Key Columns (Edit Link Template)

Modify key columns by right-clicking the link template and choosing **Edit** from the shortcut menu. You can also choose **Edit Link Template** from the **Templates** submenu of the **dbConnect** menu.

If links exist in the current drawing for a particular link template, its key columns cannot be edited. Attempting to edit a link template when links exist produces the error message shown in Figure 2.38. As the message implies, if the key columns

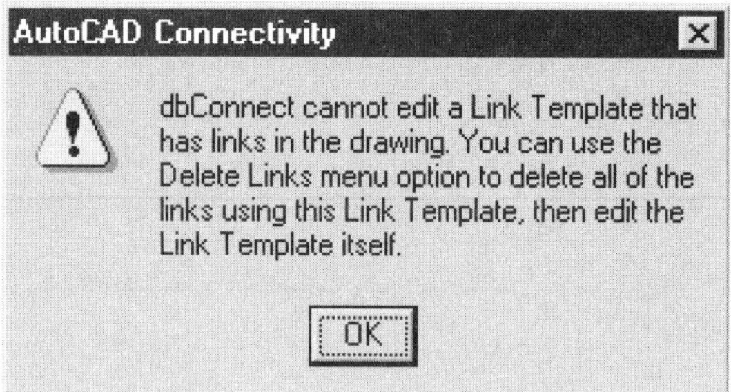

Figure 2.38 *Edit Link Template Error Message*

absolutely need to be modified, you must first delete all links associated with the link template.

DELETING A LINK TEMPLATE

Delete a link template from the drawing by right-clicking the link template and choosing **Delete** from the shortcut menu. You can also choose **Delete Link Template** from the **Templates** submenu of the **dbConnect** menu.

The same rules apply for deleting link templates as for editing them. If links exist in the current drawing for a particular link template, it cannot be deleted. Attempting to delete a link template when links exist produces the error message shown in Figure 2.39. You must delete all links associated with the link template before it can be deleted.

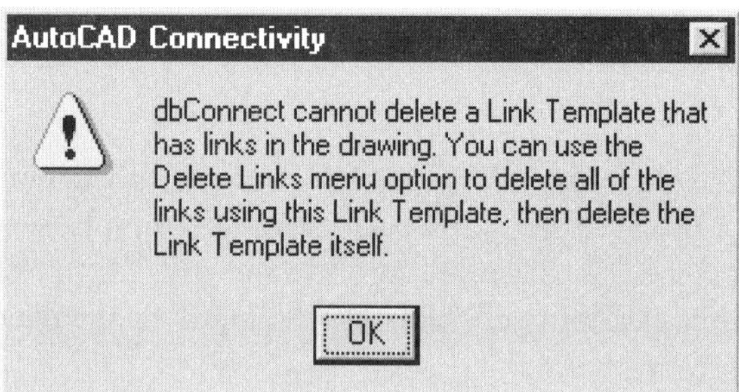

Figure 2.39 *Delete Link Template Error Message*

SHARING LINK TEMPLATES WITH OTHER DRAWINGS

At times, you may have a set of link templates in a drawing that you want to share with other drawings or other users. DbConnect provides a link template import/export capability for this purpose.

Import/Export

Issue the import and export commands by right-clicking a drawing node in the dbConnect Manager window and choosing **Import Template Set** or **Export Template Set**. These commands are also available from the **Templates** submenu of the **dbConnect** menu.

Exporting a template set stores all link templates and label templates in a file with a *.dbt* extension. This file can then be imported to another drawing. If you import a template set that contains templates that have the same name as templates already defined in the current drawing, you will be prompted to change the name, as shown in Figure 2.40.

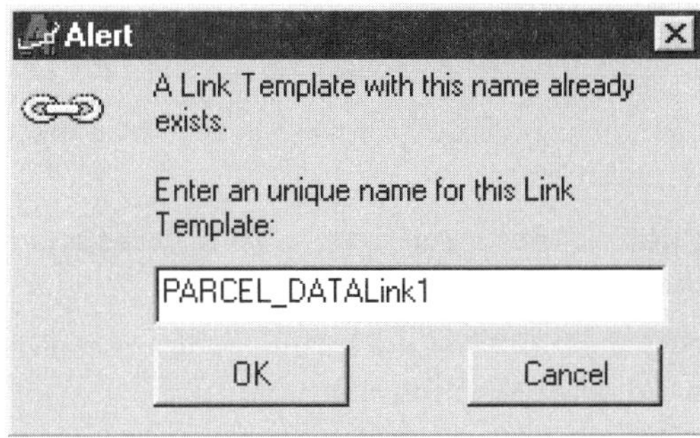

Figure 2.40 *Import Template Set Warning Message*

Drag and Drop

If you have multiple drawings open in AutoCAD, and you want to copy a template from one drawing to another, you can simply drag the template from one drawing node to another in the dbConnect Manager window.

CHECKING LINK INTEGRITY

At times, your drawing and database may become out of sync. There are a number of reasons why this may occur. For example, the database may have been modified outside AutoCAD, or link values were placed on new objects that did not match a corresponding record in the database.

Included on the CD-ROM is a copy of the Parcels database called *parcels2.mdb*, which is missing the records for Tangent Drive. In the following tutorial, we will create a data source for this database, modify the link template properties to redirect our links to the new data source, and then synchronize the links to detect the missing linked rows.

TUTORIAL 2.7 – SYNCHRONIZING LINKS

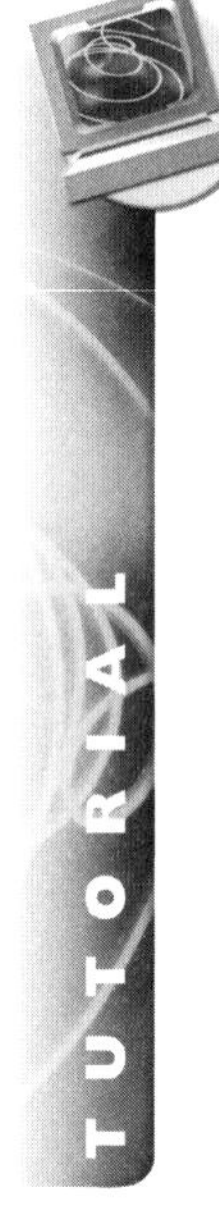

1. If you did not continue from the previous tutorial, open parcels3.dwg, display the dbConnect Manager, and connect to the parcels data source.
2. Zoom to the extents of the drawing.
3. Right-click the Data Sources branch in the dbConnect Manager and choose **Configure Data Source...** from the shortcut menu. The **Configure a Data Source** dialog box is displayed.
4. Type "parcels2" as the data source name, and click **OK** or press ENTER. This launches the **Data Link Properties** dialog box.
5. Choose **Microsoft Jet 3.51 OLE DB Provider** in the list of OLE DB Providers, and click **Next >>**.
6. Enter the full path of the parcels2.mdb file and click **OK**.
7. Double-click the parcels2 data source in the dbConnect Manager to establish the connection to it.
8. From the **dbConnect** menu, choose **Templates**, and then **Link Template Properties**.
9. Choose **PARCEL_DATALink1** from the list of available link templates (this should be the only one in the list).
10. Click **Continue**. The **Link Template Properties** dialog box is displayed.
11. In the **Data Source** pick list at the top of the dialog box, choose **parcels2**.
12. Click **OK**.
13. Right-click the **PARCEL_DATALink1** branch in the dbConnect Manager and choose **Synchronize...** from the shortcut menu. The **Synchronize** dialog box is displayed as shown in Figure 2.41.
14. At this point, you have the option to delete the bad links from the objects, or leave them. Choose **Close** to leave the links alone.

Note: If you find that the links are correct, and new rows need to be added to the table, you must add the new rows using the Data View window or some other method. New rows cannot be added from this dialog box.

15. Now we will restore the link template to its original state. From the **dbConnect** menu, choose **Templates**, and then **Link Template Properties**.
16. Choose **PARCEL_DATALink1** from the list and click **Continue**.
17. Choose **parcels** from the **Data Source** pick list
18. Click **OK**.

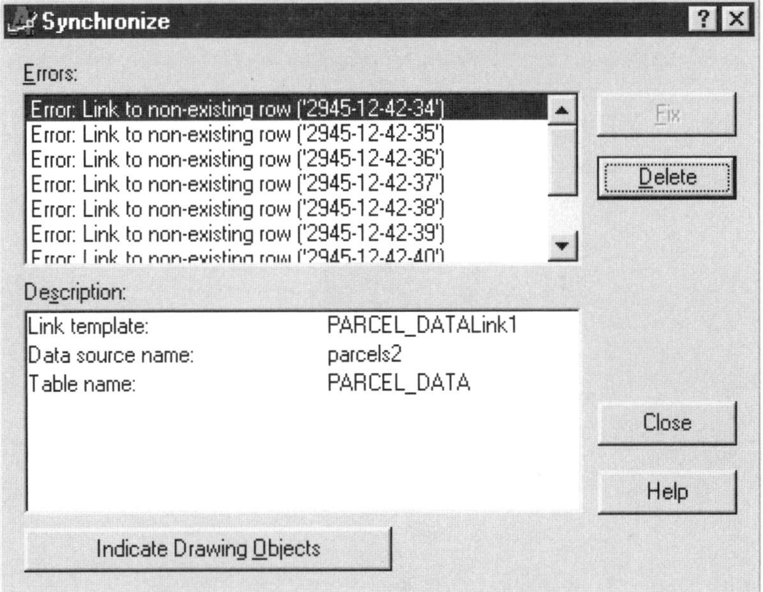

Figure 2.41 *The Synchronize Dialog Box*

Note: The parcels2 data source is no longer needed. To remove it completely, delete the parcels2.udl file from the Data Links directory.

USING OBJECT SHORTCUT MENUS

An object shortcut menu is displayed when you right-click in the drawing area when one or more graphical objects are selected. The specific items on an object shortcut menu vary depending on type of object and the number of objects selected. Typical commands you will find on an object shortcut menu include cut, copy, paste, erase, move, scale, and rotate. You can even use VBA to add your own custom menu items.

Some dbConnect functionality is available through the object shortcut menus. On the shortcut menu there are two items: **Link** and **Label**. These become available to you when either a linked object or a label is selected. Figure 2.42 shows the menu that is displayed when a single linked object is selected.

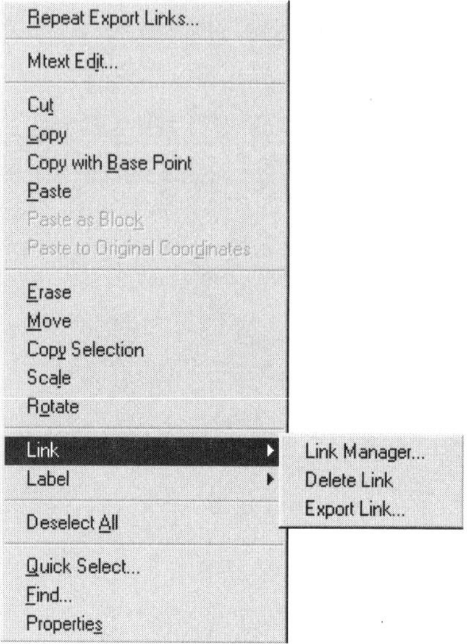

Figure 2.42 *Object Shortcut Menu for Links*

Figure 2.43 shows the shortcut menu that is displayed when a single label is selected.

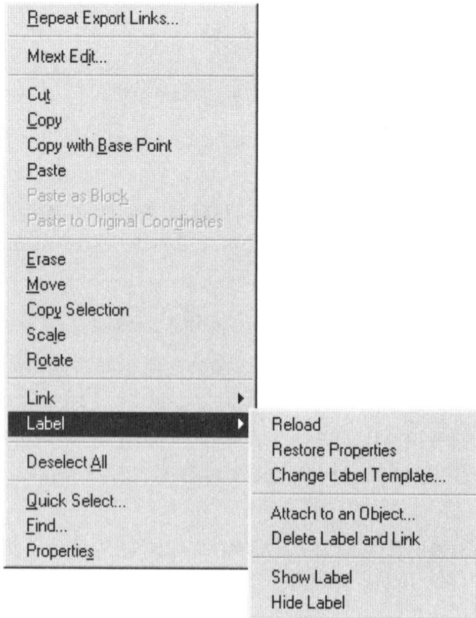

Figure 2.43 *Object Shortcut Menu for Labels*

CONVERTING OLD LINKS

If you have drawings created in earlier versions of AutoCAD that contain links to external databases, the link templates will most likely need to be updated. In earlier versions of AutoCAD, a link template was referred to as a Link Path Name (LPN). Despite the new name, they still represent exactly the same concept. The information stored in the LPN is the same, and the notation used to describe an LPN has the same structure as a link template. This notation is as follows:

Environment.Catalog.Schema.Table(LPN)

When you open an older drawing that contains links, AutoCAD 2000 automatically converts the LPN to a link template. If you have configured your data source in AutoCAD 2000 and the old LPN has the correct structure for the new environment, you're off and running. Unfortunately, in most cases the old LPN does not match the new link template in AutoCAD 2000, and it needs to be converted. DbConnect provides a link conversion utility that is made just for that purpose.

For example, if you were using ODBC to connect to an Access database in Release 14, your LPN probably looked something like this:

```
PARCELS.."c:\parcels\parcels.mdb".PARCEL_DATA↵
(PARCEL_DATALink1)
```

In AutoCAD 2000, the same environment configured through ODBC results in a link template that is slightly different:

```
PARCELS...PARCEL_DATA(PARCEL_DATALink1)
```

Included on the CD is a copy of the Parcels drawing that was created in AutoCAD Map Release 3, which is based on AutoCAD Release 14. The data source was configured through the MS Access database driver included with that release of AutoCAD Map. The old LPN for the links to the **PARCEL_DATA** table looks like this:

```
ACCESS..PARCELS.PARCEL_DATA(PARCEL_DATALINK1)
```

In the following tutorial, you will learn how to configure AutoCAD 2000 to automatically convert drawings having this LPN to the correct link template structure.

TUTORIAL 2.8 – CONVERTING OLD LINKS

1. Start AutoCAD 2000 and open parcels_r14.dwg.
2. Open the dbConnect Manager window. You should see one link template defined for this drawing called **PARCEL_DATALINK1**.
3. Connect to the parcels data source (configured in a tutorial in the previous chapter). Notice that there is a red "X" next to the link template, indicating that the environment associated with it is not connected.

4. In the dbConnect Manager, double-click the **PARCEL_DATALINK1** link template. You should get the following dialog box:

Figure 2.44 *No Data Source Connected*

5. Click **OK**.
6. From the **dbConnect** menu, select **Link Conversion**.
7. Enter the values in the dialog box as shown in Figure 2.45 and click **OK**.

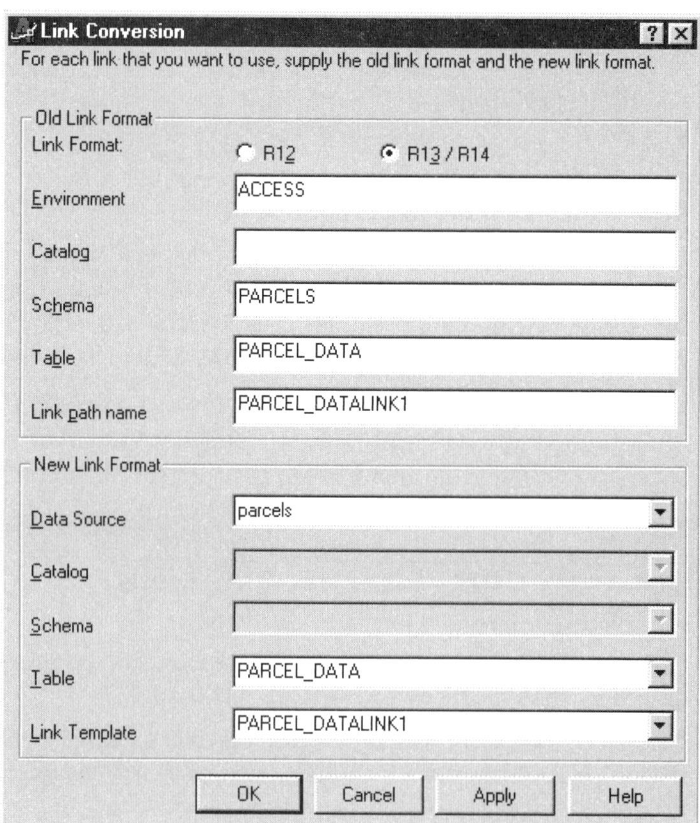

Figure 2.45 *The Link Conversion Dialog Box*

8. Close the drawing without saving it.
9. Reopen the drawing. If everything was entered correctly, you will see the **PARCEL_DATALINK1** link template, but this time the red "X" is gone.

The ASI.INI File

AutoCAD 2000 stores the link conversion information in a file called *asi.ini*, which is located in the AutoCAD installation directory. The *asi.ini* file is a text file that has entries for each conversion that AutoCAD 2000 looks for. For example, after completing the previous tutorial, your *asi.ini* file should look like this:

```
[ASE_R13]
ACCESS..PARCELS.PARCEL_DATA(PARCEL_DATALINK1)↵
= parcels...PARCEL_DATA(PARCEL_DATALINK1)
```

If you have several drawings to convert that share the same LPN information, you only have to establish the link conversion information *once* using the steps described above. As each drawing file is opened, AutoCAD 2000 will automatically update the link information.

LINKING TO OTHER DATABASE SYSTEMS

There are so many types of databases that can be linked to AutoCAD that it would be impossible to cover all of the nuances of every one of them. Up to this point, Microsoft Access has been used as the database system for all of the examples because it is so widely used. This section explains some of the issues that you will encounter when linking to other database systems. The following database systems are discussed in this section:

- Microsoft Excel
- Microsoft SQL Server
- Oracle®

The discussion on Microsoft Excel includes a series of tutorials. The files for the Microsoft Excel tutorials are included on the CD-ROM. The SQL Server and Oracle discussions are provided for information purposes and will be useful to you only if you have access to those environments.

USING MICROSOFT EXCEL

You can use a Microsoft Excel spreadsheet with dbConnect just like a database. Setting up an Excel file to be used with AutoCAD requires the following steps:

1. In the Excel spreadsheet, define a *Name* for each range of cells you want to be considered a "table" by AutoCAD.
2. Create an ODBC data source for the Excel file.

3. Create a data source in AutoCAD that refers to the ODBC data source.

The following series of tutorials takes you through the process of linking AutoCAD to an Excel spreadsheet. We will be using an Excel 97 workbook file called parcels.xls, which is simply an Excel version of the **PARCEL_DATA** table from our parcels database.

We will set up the workbook for use with AutoCAD using the steps described above. However, instead of creating a new data source for the Excel version of our parcels database, we will simply reconfigure the existing parcels data source to use the Excel ODBC data source.

The process has been divided into three tutorials for the purpose of clarity. It is recommended that you work through these as a single tutorial.

Note: These tutorials require that you have Microsoft Excel 97 and the appropriate ODBC driver installed on your system.

TUTORIAL 2.9 – DEFINING A NAMED RANGE OF CELLS IN EXCEL

1. Start Microsoft Excel and open parcels.xls. This spreadsheet is simply a copy of the **PARCEL_DATA** table from the Parcels database.
2. Press HOME to move to the top left cell in the sheet.
3. While holding down SHIFT, press CTRL + END to select the entire table.
4. From the **Insert** menu, choose **Name** and then **Define**.

Tip: As an alternative to using the **Insert** menu, you can define a new name for a selected range simply by typing the name directly in the name box, located just above the top left corner of the spreadsheet grid.

5. Type "**PARCEL_DATA**" in the **Names in Workbook** field.
6. Choose **Add.** The **PARCEL_DATA** name appears in the list (Figure 2.46).
7. Choose **Close**.
8. From the **File** menu, choose **Save**.
9. From the **File** menu, choose **Exit.** This will close Microsoft Excel.

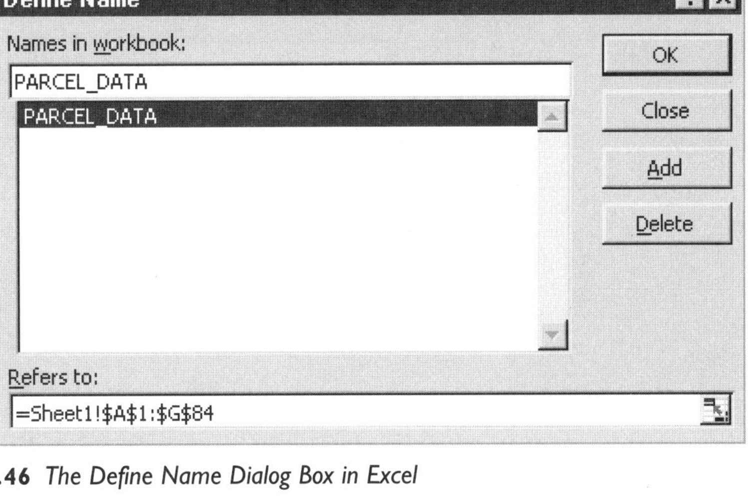

Figure 2.46 *The Define Name Dialog Box in Excel*

TUTORIAL 2.10 – CREATING THE ODBC DATA SOURCE

1. From the Windows **Start** menu, choose **Settings** and then **Control Panel**.
2. Double-click the **ODBC** icon. This launches the **ODBC Data Source Administrator**.
3. Select the **System DSN** tab, and choose **Add**. The **Create New Data Source** dialog box is displayed.
4. Choose **Microsoft Excel Driver (*.xls)** from the list of available drivers.
5. Choose **Finish**. The **ODBC Microsoft Excel Setup** dialog box is displayed (Figure 2.47).
6. Type "**PARCELS_XLS**" in the **Data Source Name** field.
7. Choose **Select Workbook....**
8. Navigate to the location of the parcels.xls file you saved earlier, and select it in the file list.
9. Make sure the **Read only** check box is not checked.
10. Choose **OK** to close the **Select Workbook** dialog box.

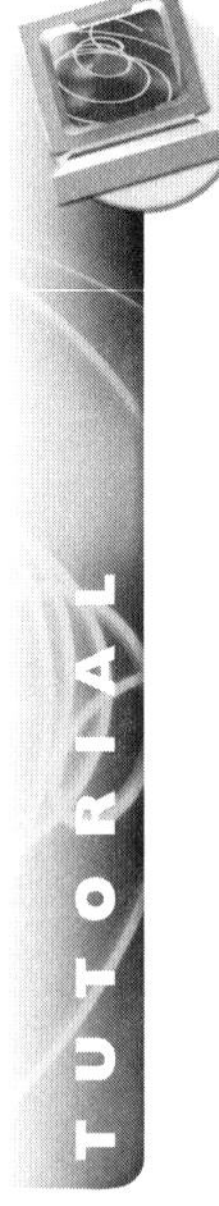

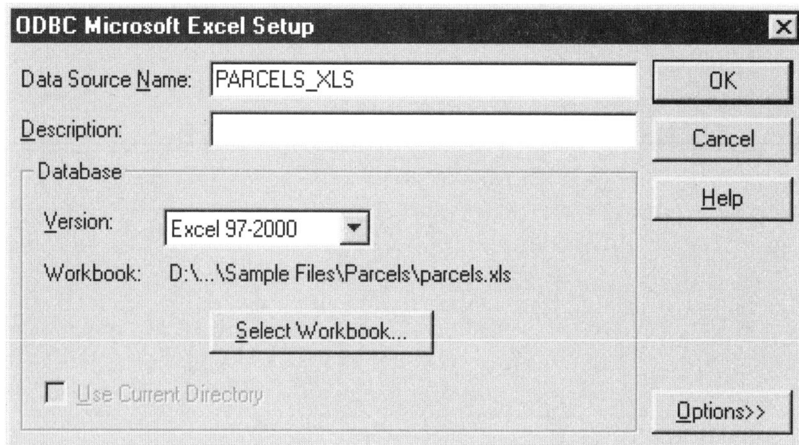

Figure 2.47 *ODBC Microsoft Excel Setup Dialog Box*

11. Choose **OK** to close the **ODBC Microsoft Excel Setup** dialog box.
12. Choose **OK** to close the **ODBC Data Source Administrator**.

TUTORIAL 2.11 – MODIFYING THE PARCELS DATA SOURCE IN AUTOCAD

1. Start AutoCAD 2000, open parcels3.dwg and display the dbConnect Manager.
2. Right-click the parcels data sources branch in the dbConnect Manager and choose **Configure...** from the shortcut menu (Figure 2.48). The **Data Link Properties** dialog box is displayed.
3. Choose the **Provider** tab.
4. Choose **Microsoft OLE DB Provider for ODBC Drivers** in the list of OLE DB Providers, and click **Next >>**.
5. Select **PARCELS_XLS** in the **Use data source name** pick list (Figure 2.49).
6. Choose **OK**.
7. Double-click the parcels node to connect to the data source. The **PARCEL_DATA** table should appear under this node.

You should now be able to use the parcels data source in much the same way as you used it when it was connected to the Access database. For example, you can use the Data View window to select linked objects in the drawing, create new links, or perform queries with the Query Editor.

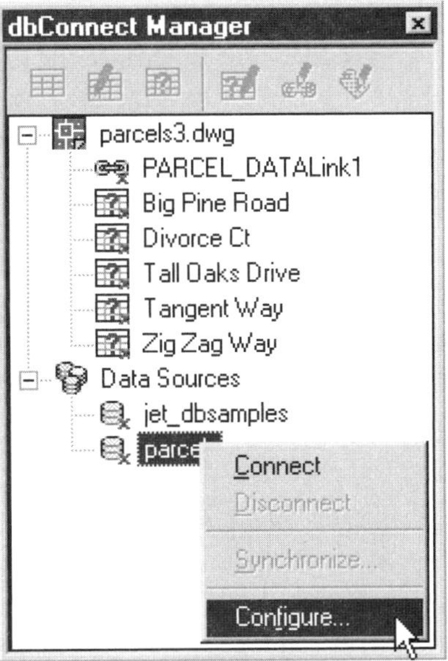

Figure 2.48 *The Data Source Shortcut Menu*

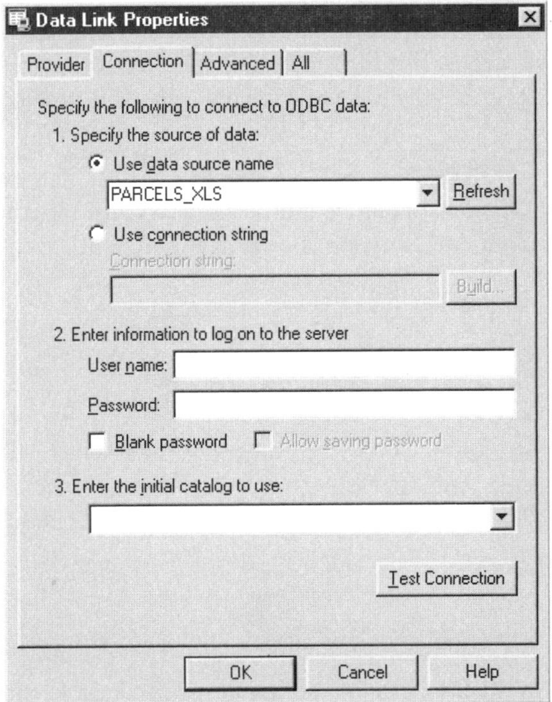

Figure 2.49 *The Data Link Properties Dialog Box*

You may notice that the Data View window is always opened in View mode, even when you explicitly select Edit mode. This is because the ODBC driver for Excel does not support updatable cursors—a capability that the data provider must support in order to allow direct modification of data in the Data View window.

MICROSOFT SQL SERVER AND ORACLE

For client/server databases, such as SQL Server and Oracle, there are very few differences in the behavior of dbConnect. Some differences you might encounter:

- When configuring a data source, you will need to know the server name and have a username and password to access databases on that server.
- You will always be prompted for a username and password the first time you connect to a data source.
- These databases support schemas, which add an additional level of nodes in the dbConnect Manager. In other words, under the data source node there are a series of schema nodes, and then each schema node contains its associated table nodes. When you log in to a client/server database, you are given specific rights to view and edit data. All of the schemas in the database may be visible in the dbConnect manager, but you will be able to view only the tables in the schemas for which your username has been granted view permission.

As you start developing applications that use client/server databases, you may encounter other subtleties. But for the most part, if you stick to standard SQL conventions, you will have relatively few problems using this type of database system.

GLOBAL DBCONNECT OPTIONS

In the AutoCAD **Options** dialog box, under the **System** tab, there are two options that pertain to dbConnect. They are shown in Figure 2.50.

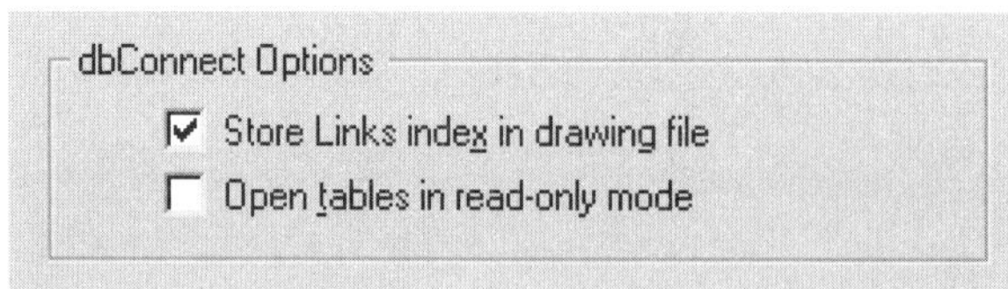

Figure 2.50 *dbConnect Options*

These two options control the following:

Store Links Index in Drawing File—Selecting this option stores a database index within the AutoCAD drawing file that enhances performance during Link Select operations. If this option is turned off, the drawing file will be smaller in size and will load more quickly.

Open Tables in Read-Only Mode—By default, when you double-click a table in the dbConnect Manager window, the table is opened in the Data View window in Edit mode. Select this option to configure AutoCAD to open tables in read-only mode by default.

SUMMARY

You now have a fairly complete picture of the dbConnect interface in AutoCAD 2000. In this chapter, you have learned how to

- Work with your linked data using the Data View window
- Create more complex queries using the Query Editor
- Build selection sets using Link Select
- Export the links of selected objects
- Modify link template properties
- Use the Synchronize feature to check link integrity
- Convert links from older drawings
- Link to Excel and other database systems

It is important to have a solid understanding of the dbConnect interface. It is the command center for AutoCAD's database connectivity engine. However, as you start working with more complex databases, your needs will quickly become more complex as well. In order to be successful with AutoCAD and databases, you need have more than just fluency with dbConnect. You need to know how to design an efficient database, how to use SQL to get meaningful information from it, and how to develop custom applications that provide users with the tools they need to manage the entire AutoCAD/database environment. The remainder of this book is dedicated to helping you develop these skills.

REVIEW QUESTIONS

1. How many different shortcut menus can be activated on the Data View window?
2. When are changes to data in the Data View window actually written to the database?
3. How do you close the Data View window without writing any of the changes to the database?
4. How can you quickly sort a single column in the Data View window?

5. What is the easiest way to get data from the Data View window into another application, such as Excel?
6. What are the differences between the Query Editor and Link Select features of dbConnect?
7. What is the difference between Edit Link Template and Link Template Properties?
8. What needs to happen before a link template can be deleted?
9. If you have two drawings open, what is the easiest way to copy a link template from one drawing to the other?
10. What utility do you use to convert links from older AutoCAD drawings?

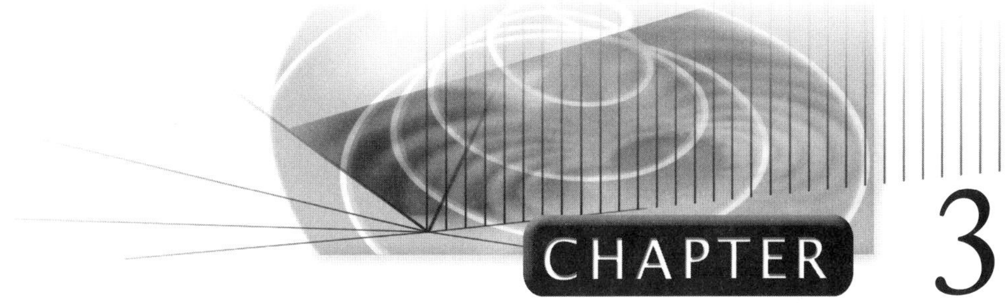

Database Design

OBJECTIVES

After completing this chapter, you will be able to

- Understand the benefits of a well-designed database
- Work through the process of designing a database
- Develop an entity relationship diagram
- Identify primary and foreign keys
- Normalize a database
- Understand referential integrity

INTRODUCTION

As AutoCAD's database connectivity features become easier to use, AutoCAD users will be more likely to venture into the world of databases. Whether it is because drawings need to be more intelligent or databases need a fancy graphic interface, databases represent a new frontier for many AutoCAD users. For example, suppose you are using AutoCAD to create a fixture layout plan for a proposed department store, and the owner wants to develop reports that show the quantity and cost of each fixture, by department. You have read the first two chapters of this book and have a working knowledge of AutoCAD and the dbConnect interface. Subsequently, you decide to create a database that is linked to the drawing and that allows you to produce the necessary reports. How do you go about designing this database?

Traditionally, as AutoCAD users, we are not exactly experienced in the art of database design. After all, designing databases is supposed to be for the information systems gurus. We have been reluctant to learn about how databases work and how they can be structured to store data efficiently. Consequently, the design of the database is often overlooked.

To many of us, when we start thinking about designing a "database," all we really want is a "table." We certainly would be happy if we could accomplish our goal with only one table in our database. A table is an easy concept to understand, and we are comfortable with it. The fact is that single-table designs just don't work, and even the simplest applications usually require multiple tables.

Database design, like any other type of design, is more of an art than a science. It is a problem-solving exercise that requires both logical and creative thinking. If you are a designer in the architectural, engineering, or mechanical industries, database design should come easy to you. You already have experience in the art of problem solving, and are likely quite passionate about it within your industry. Whether you are designing a building, a highway, a piece of machinery, or a database, it is fundamentally a problem-solving process. In addition, just as in designing real-world things, each new design must be approached as an entirely new set of challenges.

As we start to venture out of the confines of the AutoCAD graphic database and into the world of relational databases, we must be armed with a basic knowledge of how databases are designed. This chapter takes you through a structured process of designing databases and uses two examples to illustrate this process. The first example is relatively simple and will give you an overview of the process. The second example is more complex, and it will be worked through in more detail. The finalized design will then be used in our discussion of SQL in the next chapter.

This chapter focuses on database design, independent of any AutoCAD linking issues. In Chapter 5 we will revisit the design process and discuss the issues that are introduced by the integration of databases and AutoCAD.

WHY IS IT SO IMPORTANT?

How you organize the data in your external database is critical to the success of your application. Whether you are using a graphic interface to manage and represent information in your database or not, a well-thought-out design is essential. You must take the time to determine exactly how your data will be organized, before you start linking your database to AutoCAD.

AutoCAD imposes very few restrictions on how a database is linked to your drawing and has no problem linking to a poorly designed database. Linking your drawing to the database is the easy part. It is when you start maintaining the data and performing queries that a poorly designed database begins to rear its ugly head. Soon, you become frustrated with AutoCAD because of its "limitations" when it works with your database. However, closer examination in these cases usually reveals that poor database design—not the AutoCAD software—is the culprit.

A poorly designed database can not only result in limitations in how information can be retrieved from the database, it can also make maintenance of the database very difficult, which can lead to inconsistent and inaccurate information. The structure of your database is the key to its accuracy and integrity. A database will not be maintained properly if the design makes it difficult or inefficient. Once the users of the data lose faith in the accuracy and completeness of the data, the database will eventually become obsolete.

HOW DATABASES ARE USED

When an external database is used to store critical information about an application, the application must be able to perform a variety of operations on the database, including

- Populating the database with new information
- Changing existing information
- Removing information
- Producing reports

The first three operations should go without saying. They represent the general maintenance activity that is necessary for any database. The difficulty or ease with which the database performs these kinds of operations is called the *maintenance effort*. The fourth task—producing reports—is where the database really does its work. The term "producing reports" is a generic term used to describe any operation that pulls data out of the database and allows you to look at it in different ways—it is the process by which raw *data* becomes *information*.

The success of a database design can be measured by its ability to turn data into valuable information. In order to present your data in different ways, you need to be able to formulate complex queries on the data, and the database needs to be flexible enough to respond to these queries.

GOALS OF DATABASE DESIGN

The goals of a good database design are as follows:

Eliminate redundancy—A database that contains redundant data is difficult to maintain and eventually results in inaccurate or incomplete information.

Reduce maintenance effort—The maintenance effort is the ease with which the database can handle maintenance operations such as inserting, updating, or deleting rows. Eliminating redundancy is one of the best ways to reduce maintenance effort because data that exists in the database will only exist in one place and can typically be managed with a single operation.

Maintain data integrity—The database design should support referential integrity, as well as make use of the table and field integrity features of the database system.

- **Allow for growth**—A well-designed database is one that will grow with the needs of the organization, without having to be completely redesigned.

THE RELATIONAL DATA MODEL

The relational model for databases that is in use today was originally developed in the late 1960s by Dr. Edgar F. Codd. Codd introduced a model for organizing information by grouping information into *relations* (tables) and establishing relationships between the tables. Many of the concepts you will learn in this chapter, and throughout this book, are based on this model.

ENTITY RELATIONSHIP (ER) DIAGRAMS

In this chapter, we will approach the database design process using a data modeling method called entity relationship (ER) diagramming. An *entity relationship diagram* (ERD) identifies and communicates the various entities (tables) in your design and shows the relationships that exist between those entities. As the design matures, the ERD can be expanded to show the specific attributes (fields) associated with each entity. When the design is complete, the entities and relationships are modeled in a relational database system.

COMPONENTS

Figure 3.1 shows a simple example of an entity relationship diagram. The three primary components of an ERD are entities, attributes, and relationships. A description of each component follows.

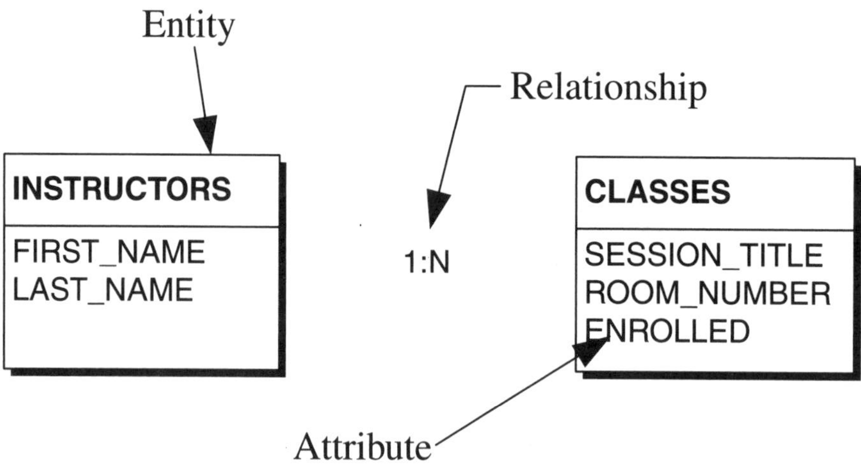

Figure 3.1 *Components of an Entity Relationship Diagram*

Entities

An entity represents a single logical object that exists in the model. It encapsulates all of the attributes related to that object. In a database, a table holds all the information related to a specific type of entity. In Figure 3.1 there are two entities: instructors and classes.

Attributes

Attributes hold the specific information that describes an entity. In a database, attributes are represented as fields (or columns) in a table. In the example, instructors are described by their first name and last name. In other words, first name and last name are said to be *attributes* of an instructor.

Relationships

A relationship describes an association or dependency that exists between two entities. Our sample diagram shows a *one-to-many* relationship between instructors and classes. In plain English, we are saying "An instructor teaches one or more classes." In other words, an instructor can teach one or more classes, but each class can only have one instructor.

Note: This diagram and the diagrams you will see throughout this book use a "crow's foot" method of depicting relationships. The "one" end of the line is a single line, and the "many" end of the line looks like the three claws of a crow's foot.

WEAK ENTITIES

A *weak entity* is an entity whose existence depends on the existence of another entity. In our example, "classes" is a weak entity because you cannot have a class that does not have an instructor assigned to it. If an instructor is removed from the database, then the class must also be removed or assigned to a different instructor. This kind of rule can be imposed at the database level. We will discuss this concept in more detail later in this chapter.

THE DESIGN PROCESS

While there are good designs and bad designs, there is no single right or wrong way to design a database. The merits of a particular database design can only be measured by how effective the database is at solving the specific problems of the application.

Developing an ER diagram is only one part of the process. The ER diagram is a tool that helps you visualize the design, which is entirely based on the needs of the application. To fully understand the needs of the application, you should start by interviewing the people who will be using the application and involve those users throughout the entire design process.

ESTABLISHING USER REQUIREMENTS

Establishing a detailed list of user requirements for the database is probably the single most important stage of the design process. An understanding of how the database will be used gives you a head start into the database design. A documented set of requirements becomes the foundation upon which all decisions are made during the design process.

In addition, having a good set of documented requirements keeps you focused on the real issues. It helps you avoid doing things in your database design just because you did them in a previous database design. It also keeps you from over-designing the database and creating a database with more complexity than is really necessary for the application.

You must consider how the data will be used before you can start designing. Here are some things to think about:

- Who are the users of the database?
- What types of information will be tracked?
- How will the data get *into* the database?
- What kind of information will users want to get out of the database?
- What types of reports will be needed?
- What are the dynamics of the data (how often do certain data change)?

The Interview Process

The best way to establish a solid foundation of requirements is to interview the key users of the database. Go to the people who will actually be using the database, and spend time talking about how they intend to use the data. Understand the nature of their business; where the data come from, who maintains the data, and who uses the data.

At the beginning of the process, you should perform some initial interviews to get the general information you need to identify the entities. This gives you an opportunity to develop a "first draft" design. Then, as the design process continues, you should keep the users involved throughout the entire design process.

The interviews themselves should be with just one or two people from a single department or group within the organization. This helps keep the conversation focused on a specific set of issues. If you involve too many people from too many groups, some people will invariably dominate the conversation, thus preventing other important issues or viewpoints from being heard.

Also, make sure you take good notes. If you are interviewing a large organization with many groups involved, you will be performing many interviews. You should not rely on memory alone as you start developing the design. Your notes will help you organize and categorize the information you gathered during the interview process.

While each interview will take its own course, you should try to have a series of questions that you ask during every interview. These questions are usually very general, open-ended questions, such as "What are the top three factors that are critical to the success of your group or department?" This makes it easier to compile, prioritize, and summarize the interview results.

The Problem Statement

The first part of the needs assessment process is to create a simple statement, of one or two sentences, that concisely defines the purpose of the database. This statement is usually developed as a result of an initial interview with the chief sponsor of the project. This person is usually a director or manager of a company or department for which the database is being designed. The chief sponsor will provide the best view of the "big picture." This interview will help you understand the nature of the organization and how the sponsor envisions the database helping the organization accomplish its goals.

Let's look at the example that was mentioned earlier in this chapter. Suppose you are asked by a department store owner to develop a database to help him manage his assets. You conduct an interview and develop the following problem statement:

The purpose of the database is to help project managers keep up with the quantity and cost of fixtures during the design process of a department store.

This statement is very concise and to the point. No matter how complex the problem is, the problem statement should summarize the goals in just one or two sentences. The more complex the problem, the more general the statement. As we proceed with the design process, we will break down the problem statement into specific tasks or requirements of the system. Remember that the problem statement is not used to define the database—it is used to define the problem.

Note: Keep in mind that we are trying to illustrate the design process using a very simplified example. Most problems will be much more complex, and therefore the problem statement will be more generic.

IDENTIFYING AND DEFINING ENTITIES

The next step is to identify the entities that will make up your database design. One way to get a quick, preliminary view of the necessary entities is to simply look at the problem statement and extract the nouns (objects) from the statement.

Our problem statement describes three distinct entities: departments, fixtures, and stores. This is by no means a complete set of entities, nor does it constitute a complete design. It does however, give you a good starting point.

Sometimes your problem statement will contain nouns that represent attributes, rather than entities. Make sure you can differentiate between an entity and an attribute in your problem statement. If your problem statement contains too many attributes, it may be getting too specific. For example, *quantity* and *cost* are nouns in our problem statement that are attributes, and not entities. We could probably further generalize our problem statement to eliminate this kind of detail. An example is as follows:

The purpose of the database is to help project managers control all aspects of the design process of a department store.

Then, you can get detailed information about the nature of the data by asking more questions during the initial interview. For example:

> Q. What are the responsibilities of the project manager?
>
> A. The project manager is responsible for ensuring that projects stay on schedule and within budget.
>
> Q. What critical information do they need to help estimate costs?
>
> A. They need to know the amount of space being allocated to each department during design and the quantity and cost of each fixture as the sales areas are being laid out.

With just a few questions, we have a great deal of additional information that will be considered in our database design. For now, we will begin developing an ER diagram that includes the following three entities:

- Stores
- Departments
- Fixtures

IDENTIFYING RELATIONSHIPS

The next step is to consider each pair of entities and determine if any type of relationship exists. For example, consider the two entities store and department. A relationship exists between these two entities because a store *contains* departments. There is also a relationship between stores and fixtures, because a store *contains* fixtures. The relationship between departments and fixtures is dependent on the needs of the application. If fixtures never need to be classified by department, then an explicit relationship may not be necessary. On the other hand, if a fixture *is* assigned to a department, then the relationship is necessary.

Figure 3.2 shows an initial ER diagram for the department store example.

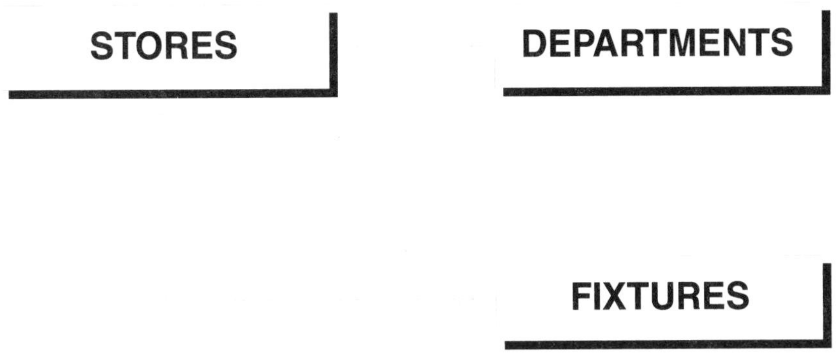

STORES

DEPARTMENTS

FIXTURES

Figure 3.2 *Department Store ER Diagram*

In this model, we could say that departments and fixtures are weak entities because a fixture cannot exist without a department, and a department cannot exist without a store. If this is the case, then the relationship between stores and fixtures is redundant. Although it is true that a store contains fixtures, that relationship is already implied by the relationship between stores and departments and between departments and fixtures.

In a database, these redundant relationships can be eliminated because you can always determine which fixtures belong to which stores using the other relationships.

IDENTIFYING ATTRIBUTES

The next step is to take each entity and determine the attributes that will be used to describe that entity. In this stage of the design process, you should conduct interviews with the users of the database. Suppose, for example, we determine that stores have a name and a location, departments have a name, and fixtures have a fixture type and a cost. Figure 3.3 shows the ER diagram with the attributes shown.

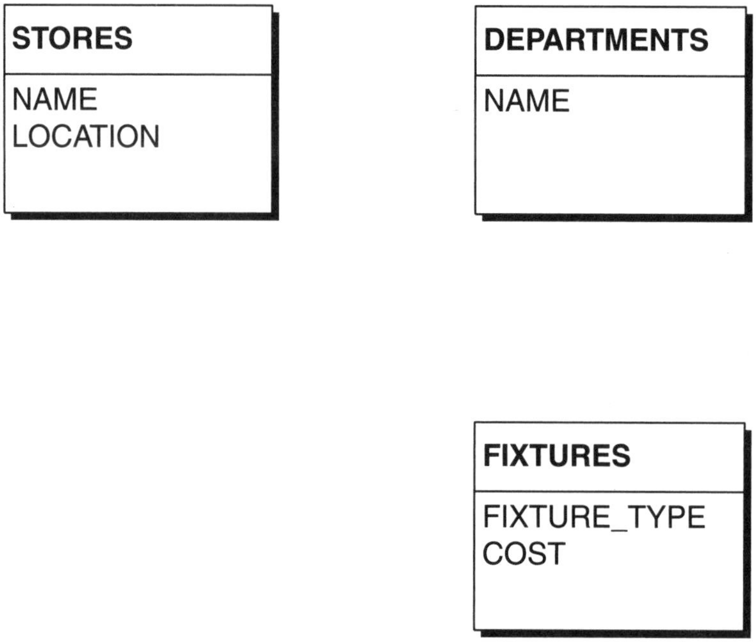

Figure 3.3 *Department Store ER Diagram with Attributes*

IDENTIFYING PRIMARY AND FOREIGN KEYS

A field in a table that is used to uniquely identify each row is called a *primary key*. A *foreign key* is a field used to establish a relationship between two tables. The foreign key in one table typically contains a value that exists in the primary key of the other table.

Primary Keys

In order for a field to be considered a true primary key, it must comply with the following rules:

- There can only be one primary key in a table.
- It must be unique within the table.
- It must not contain null values.
- Its value is rarely modified, if ever.

It is always good practice to use primary keys, even if they are not referred to by any other table. A primary key can be a field that exists as part of the original set of fields in your design, or it can be a separate field that exists solely for the purposes of acting as the key.

In our department store database, we will add a field to each table to be used as the primary key. Figure 3.4 shows the entity relationship diagram with the primary keys shown in italics.

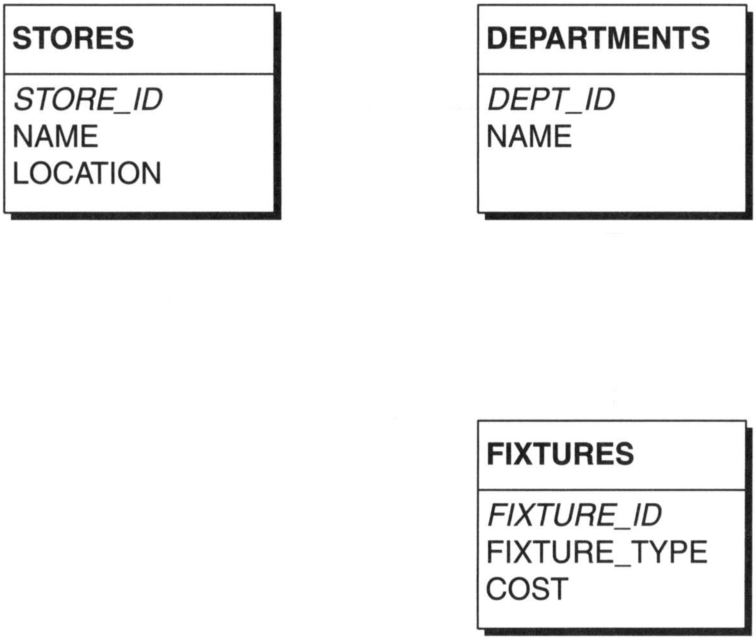

Figure 3.4 *ER Diagram with Primary Keys Identified*

Foreign Keys

A foreign key is necessary in any table that is on the "many" side of a one-to-many relationship. For example, since a store contains multiple departments, each department must somehow identify the store to which it belongs. For this to be accomplished, an additional field is required in the department table. Each row in the department table has a value in this field that matches the primary key value of its associated row in the store table. This field is called a foreign key.

Figure 3.5 shows an example of how primary keys and foreign keys relate to each other.

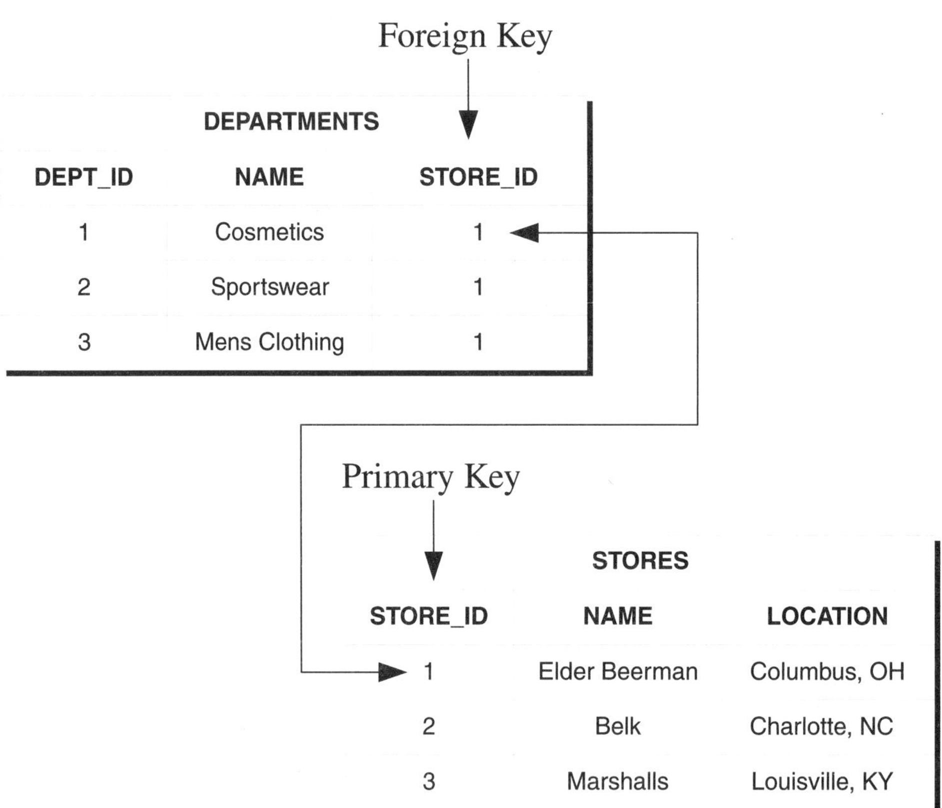

DEPARTMENTS

DEPT_ID	NAME	STORE_ID
1	Cosmetics	1
2	Sportswear	1
3	Mens Clothing	1

STORES

STORE_ID	NAME	LOCATION
1	Elder Beerman	Columbus, OH
2	Belk	Charlotte, NC
3	Marshalls	Louisville, KY

Figure 3.5 *Primary and Foreign Key Relationships*

While a table can only have one primary key, it can however, have multiple foreign keys. With the exception of the stores table, each of the tables in our database is on the "many" side of a one-to-many relationship, and foreign keys must be established. Figure 3.6 shows the database model with the foreign keys, which are identified by a "(FK)" following the field name.

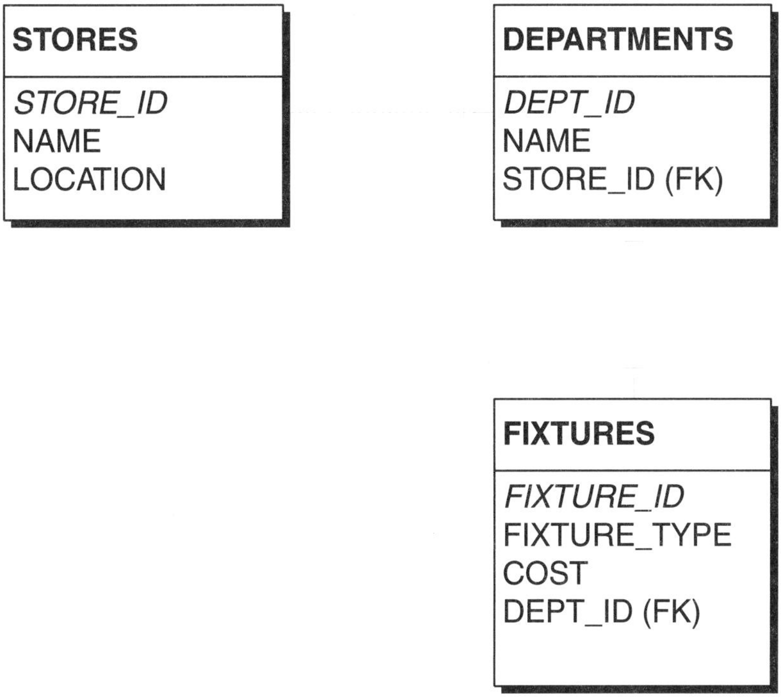

STORES
STORE_ID NAME LOCATION

DEPARTMENTS
DEPT_ID NAME STORE_ID (FK)

FIXTURES
FIXTURE_ID FIXTURE_TYPE COST DEPT_ID (FK)

Figure 3.6 *ER Diagram with Primary and Foreign Keys Identified*

NORMALIZATION

The final step in the design process is called *normalization*. Normalization is a refinement process that strives for the most efficient way to store the data in your database. The primary purpose of normalization is to eliminate redundant or duplicate data. As you work through this refinement process, it is often necessary to create additional tables (entities) and, in turn, additional relationships.

During the normalization process, the database design is carefully examined through a series of rules called *normal forms*. Several normal forms are in use in the database industry. To keep things simple, our discussion on normal forms will be limited to *First Normal Form*, *Second Normal Form*, and *Third Normal Form*. These represent the most commonly used normal forms and are suitable for most database designs.

First Normal Form

To comply with the first normal form, a table must not contain any repeating groups. A repeating group is a series of fields that attempt to define more than one instance of a single type of object. Consider the department store database design shown in Figure 3.6. The design as it stands now satisfies first normal form, because no such repeating groups exist.

Let's introduce a condition that will help illustrate the problem that first normal form is meant to solve. Suppose we need to track the area that each department occupies on each floor of the store. We might be tempted to simply modify the **DEPARTMENTS** table to include that information, as shown in Figure 3.7.

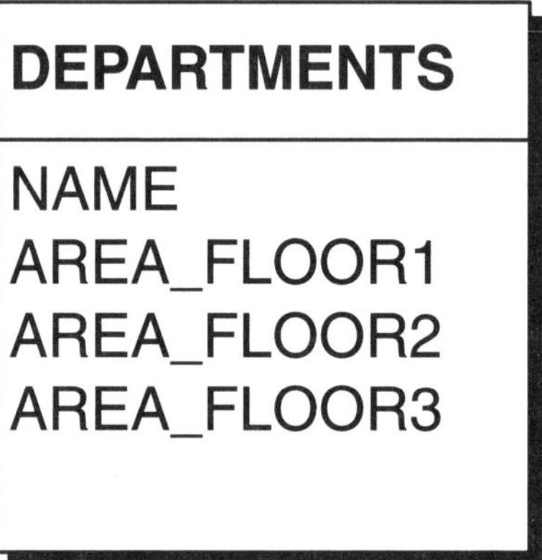

Figure 3.7 *Modified DEPARTMENTS Table*

Doing this, however, violates first normal form because there is a group of 1...*n* attributes called **AREA_FLOOR**. This type of repeating group can seriously limit the growth potential of a database. This **DEPARTMENTS** table, for example, assumes that a store will never have more than three floors. It also does not give us the flexibility to further categorize "instances" of departments other than by floor.

To solve this problem, and bring our database to first normal form, we must move the area information into a separate table. This table, which we call **DEPT_AREA**, tracks specific "instances" of a department and indicates the floor on which each instance exists. Figure 3.8 shows the modified database design, which now satisfies first normal form.

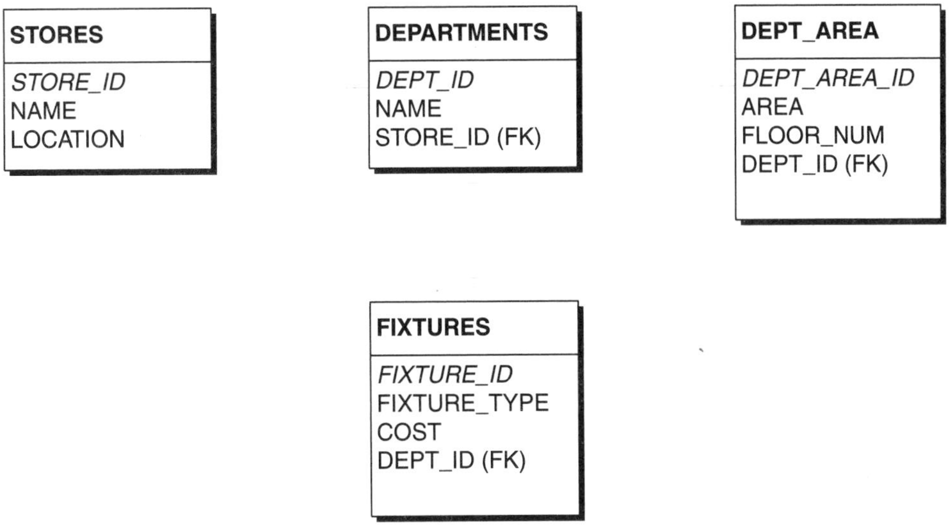

STORES
STORE_ID
NAME
LOCATION

DEPARTMENTS
DEPT_ID
NAME
STORE_ID (FK)

DEPT_AREA
DEPT_AREA_ID
AREA
FLOOR_NUM
DEPT_ID (FK)

FIXTURES
FIXTURE_ID
FIXTURE_TYPE
COST
DEPT_ID (FK)

Figure 3.8 *Department Store Database in First Normal Form*

Getting past first normal form is relatively easy. In fact, if you have properly identified the entities required for your database, you have most likely already reached first normal form. However, it still needs to be part of the normalization process, ensuring that your database is ready to be tested against the next set of criteria imposed by second normal form.

Second Normal Form

To bring our database to second normal form, we must eliminate the possibility of redundant or duplicate data within our database. To help identify potential areas in our design where duplication might occur, we need to start thinking about the actual *data* that will be contained within our tables.

A potential area where duplication is likely to occur is in any attribute that defines a "type" for the entity to which it belongs. In our example, we have an attribute that belongs to our fixtures entity that describes the type of fixture. When an entity becomes a populated table in our database, this type field could potentially have multiple instances of the same type. In addition, if the "cost" attribute is associated with a particular type (rather than an instance of the type) it will also contain duplicate data.

For this issue to be resolved, there needs to be a distinction between an instance of a fixture and a fixture type. We should move information about the "type" of fixture to a separate table and then establish a relationship between the two tables. The modified design is shown in Figure 3.9.

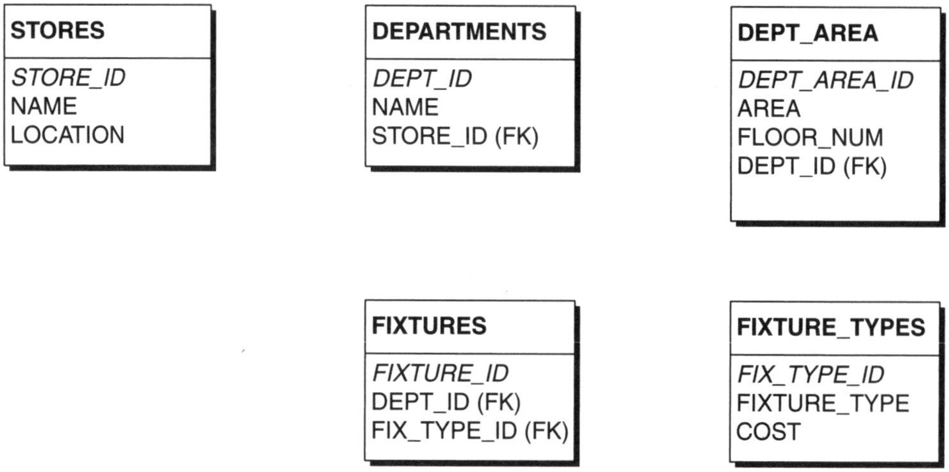

STORES
STORE_ID
NAME
LOCATION

DEPARTMENTS
DEPT_ID
NAME
STORE_ID (FK)

DEPT_AREA
DEPT_AREA_ID
AREA
FLOOR_NUM
DEPT_ID (FK)

FIXTURES
FIXTURE_ID
DEPT_ID (FK)
FIX_TYPE_ID (FK)

FIXTURE_TYPES
FIX_TYPE_ID
FIXTURE_TYPE
COST

Figure 3.9 *Design in Second Normal Form*

Third Normal Form

Third normal form imposes the following rule: All the fields in a given table should be directly related to the primary key of that table. In other words, each field in the table must exist to help describe the kind of object that the table represents.

One condition in our department store database that violates this rule is the **FLOOR_NUM** field in the **DEPT_AREA** table. The purpose of this table is to track instances of a department and the area associated with each instance. A floor is logically a type of object separate from a department instance.

Also, by placing the floor field in this table, we lose control over what floor numbers can be used for each instance of a department. In other words, the database cannot prevent a value of "4" being entered here, even if the store does not have four floors.

To address this problem, we should add a **FLOORS** table that contains a record for each floor for each store. This may result in a small amount of data duplication within the **FLOORS** table, but it gives the control over the actual number of floors that exist within each store. If we place this table in our design between the **STORES** table and the **DEPARTMENTS** table, the database can enforce more strict rules about the existence of departments.

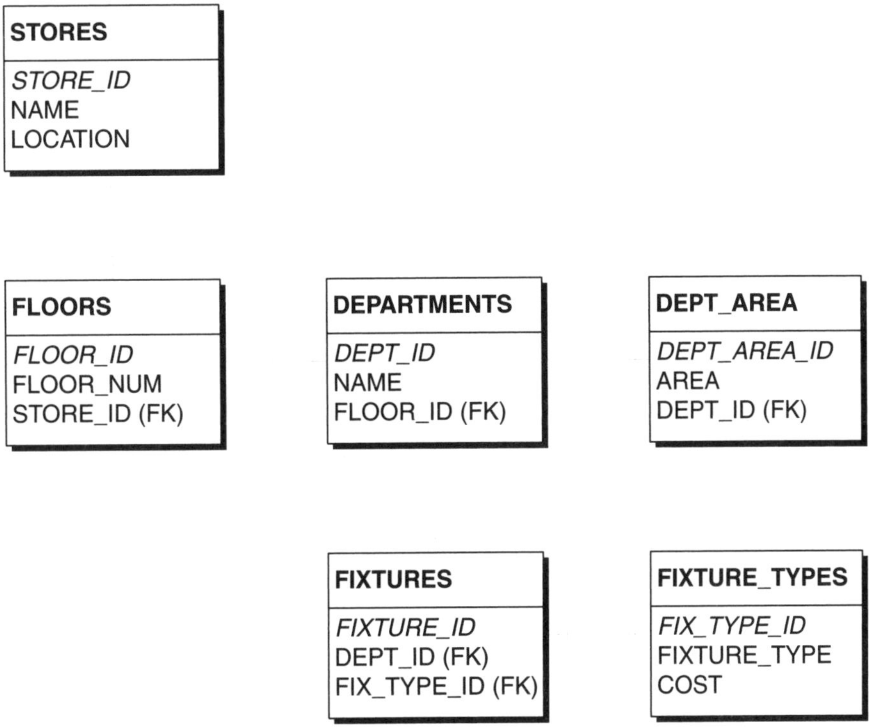

STORES
STORE_ID
NAME
LOCATION

FLOORS
FLOOR_ID
FLOOR_NUM
STORE_ID (FK)

DEPARTMENTS
DEPT_ID
NAME
FLOOR_ID (FK)

DEPT_AREA
DEPT_AREA_ID
AREA
DEPT_ID (FK)

FIXTURES
FIXTURE_ID
DEPT_ID (FK)
FIX_TYPE_ID (FK)

FIXTURE_TYPES
FIX_TYPE_ID
FIXTURE_TYPE
COST

Figure 3.10 *Database Design with Floors Table*

At closer examination of the new design in Figure 3.10, you will find a major flaw in how the floor problem was handled. Even though the design satisfies third normal form, our design now violates second normal form. The **DEPARTMENTS** table represents information related to a department, and the **DEPT_AREA** table represents information about an *instance* of a department. Logically speaking, a floor does not contain departments—a floor contains department instances, and each instance is associated with a specific department.

With the design shown in Figure 3.10, if a department exists on more than one floor, then its name must exist more than once in the **DEPARTMENTS** table. What we need to do is switch the **DEPARTMENTS** and **DEPT_AREA** tables in the design. That way, each department name will only exist once in the **DEPARTMENTS** table, regardless of how many stores, floors, or department instances exist. Figure 3.11 shows our final database design. Notice also that the **FIXTURES** table is now related to the **DEPT_AREA** table, since a fixture must exist within a department instance, rather than be assigned directly to a department.

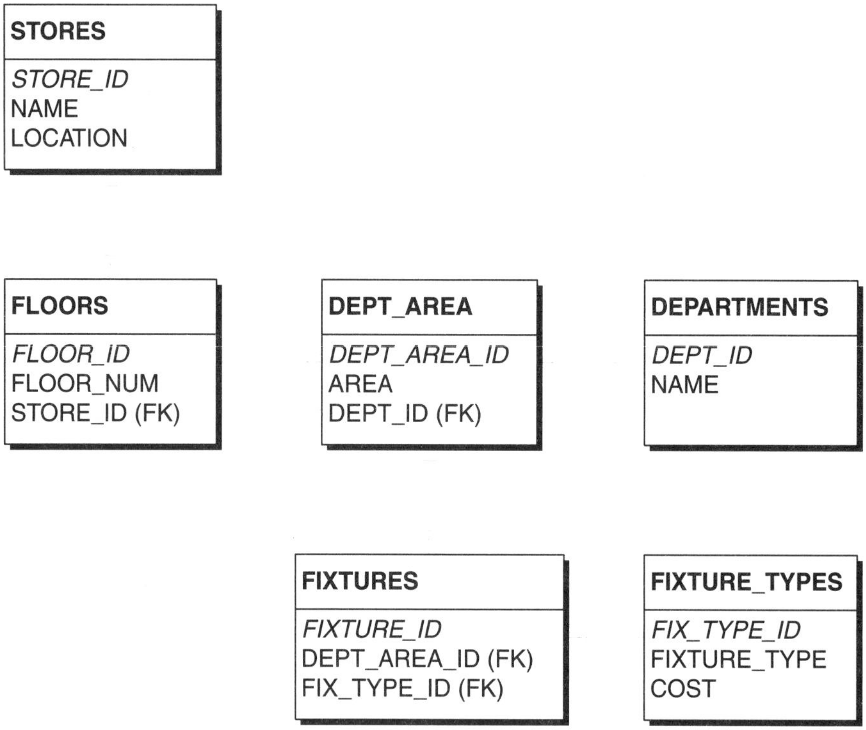

STORES
STORE_ID
NAME
LOCATION

FLOORS
FLOOR_ID
FLOOR_NUM
STORE_ID (FK)

DEPT_AREA
DEPT_AREA_ID
AREA
DEPT_ID (FK)

DEPARTMENTS
DEPT_ID
NAME

FIXTURES
FIXTURE_ID
DEPT_AREA_ID (FK)
FIX_TYPE_ID (FK)

FIXTURE_TYPES
FIX_TYPE_ID
FIXTURE_TYPE
COST

Figure 3.11 *Final Department Store Design*

DESIGNING THE CONFERENCE DATABASE

Now that you have an idea of how the process works, let's look at a slightly more complex example. In this example, you have been asked by a conference director to design a database to help manage the classes, speakers, and enrollment for a conference. We will work through the same process as the previous example and finish the chapter with a completed database design. This database will then be used throughout the next chapter, when we introduce the query language, SQL.

ESTABLISHING USER REQUIREMENTS

First, we will conduct an interview with the conference director to determine the basic requirements of the system and develop a problem statement. We should start our interview with an open-ended question regarding the role of conference director, as follows:

> Q. What responsibilities do you have as director of this conference?
>
> A. I am responsible for recruiting speakers, gathering session proposals, and overseeing the development of the conference program.

Next, we need to understand what other roles exist, so we can start establishing a list of other potential users of the database.

> Q. What are the specific roles of the various members of your team?
>
> A. There are three primary positions: the conference director, the program director, and the registration director.
>
> Q. What are the responsibilities of the program director?
>
> A. The program director is responsible for developing the conference program. This includes developing the conference schedule, assigning classes to specific rooms and timeslots.
>
> Q. What are the responsibilities of the registration director?
>
> A. The registration director is responsible for maintaining the list of attendees at the conference and keeping up with the classes that each attendee signs up for.
>
> Q. So the purpose of this database is to manage all the activity related to the planning, registration, and execution of the conference. Correct?
>
> A. That is correct.

At the completion of this interview, we have a problem statement, which is as follows:

The purpose of this database is to manage all the activity related to the planning, registration, and execution of the conference.

IDENTIFYING AND DEFINING ENTITIES

As we look at our notes from the initial interview, we can start to identify some of the entities (the nouns in the text) that will make up the design of our database. These are:

- Speakers
- Classes
- Rooms
- Timeslots
- Attendees

IDENTIFYING RELATIONSHIPS

The next task is to define the relationships that exist between the various entities in the design. In this example, the following relationships exist:

> **Classes and Speakers (one-to-many)**—A speaker teaches one or more sessions, but each session is taught by only one speaker.
>
> **Classes and Rooms (one-to-many)**—A room holds multiple classes during the course of the conference, but each class will be held in only one room.

Classes and Timeslots (one-to-many)—During any particular timeslot, there will be multiple classes being held, but each class is held during a single timeslot.

Classes and Attendees (many-to-many)—A class has multiple attendees, and each attendee can sign up for multiple classes.

Figure 3.12 shows the initial entity relationship diagram for the conference database.

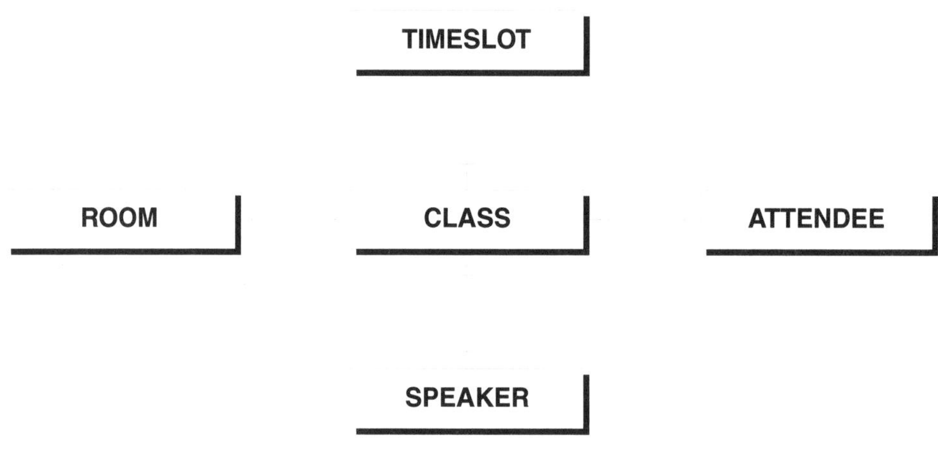

Figure 3.12 *Initial ER Diagram for the Conference Database*

IDENTIFYING ATTRIBUTES

The next step is to determine what attributes are necessary to define each entity. Again, it is a good idea to interview the key users of the database to find out what kinds of information they need to track in the database and what types of queries and reports are needed.

Note: For the purposes of this example, the database is greatly simplified. The primary focus of this exercise is to understand the process. The database design is lacking some entity-specific information, which should not detract from the presentation of database design concepts.

The following is a list of the attributes we will incorporate into this database.

Speakers

The speakers entity will contain the following attributes:

- First name
- Last name

Classes

The classes entity will contain the following attributes:

- The title of the session
- The general subject category, or "track"
- The type of class (lab or lecture)
- The grade level (beginner, intermediate, or advanced)

Rooms

The rooms entity will contain the following attributes:

- Room number
- Maximum capacity of the room
- The type of room (lab or lecture)

Timeslots

The timeslots entity will contain the following attribute:

- A text description containing the day and time

Attendees

The attendees entity will contain the following attributes:

- First name
- Last name

Figure 3.13 shows the current ER diagram with the attributes shown.

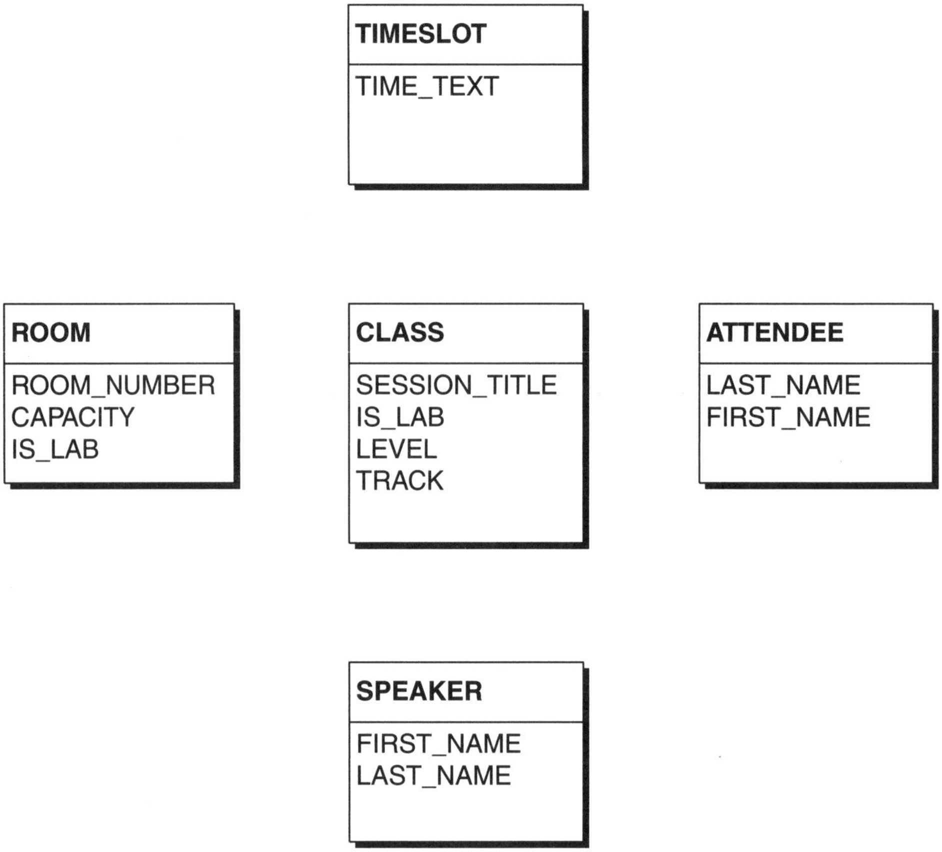

Figure 3.13 *Conference ER Diagram with Attributes*

IDENTIFYING PRIMARY AND FOREIGN KEYS

Handling the One-to-Many Relationships

In our earlier example of the department store, we learned that each one-to-many relationship can be handled with a primary key/foreign key relationship. Each table on the "one" side of the relationship has a primary key, and each table on the "many" side of the relationship has a foreign key that associates it with the appropriate row in the related table. In the conference database, we need a primary key for each of the **TIMESLOT**, **ROOM**, and **SPEAKER** tables. Then in the **CLASS** table, we need three foreign keys that establish the relationships to the other tables.

Handling the Many-to-Many Relationships

Handling a many-to-many relationship is a bit more complicated. At first, you might consider adding a foreign key to the **ATTENDEE** table that is associated with the primary key in the **CLASS** table. The problem with this scenario is that it would require duplication of the attendee name for every class for which the attendee was signed up. This kind of duplication would violate second normal form and make the database very difficult to maintain.

The better approach is to create a *linking* table. A linking table is a table that is used solely for the purposes of establishing a many-to-many relationship between two other tables. It generally contains two foreign keys—one for each table. Figure 3.14 illustrates the relationships between the linking table and the **CLASS** and **ATTENDEE** tables.

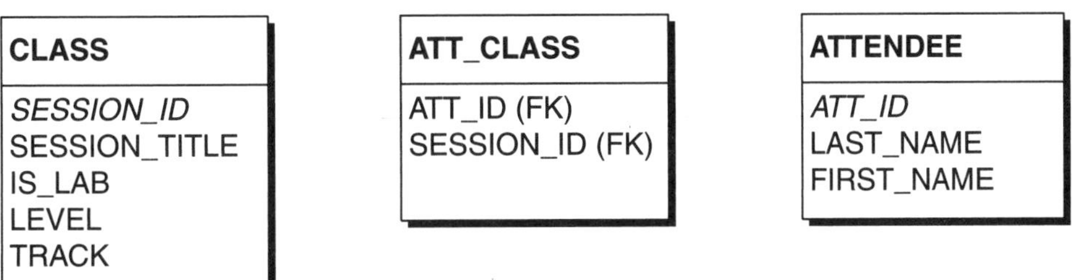

Figure 3.14 *The Linking Table*

When you create a linking table, you should give it a name that describes its purpose. Here, we have called the linking table "**ATT_CLASS**" to indicate that it defines the relationship between attendees and classes. As you can see in Figure 3.14, the linking table takes the many-to-many relationship and splits it into two one-to-many relationships. There is a one-to-many relationship between the **ATT_CLASS** table and the **ATTENDEE** table, and there is a one-to-many relationship between the **ATT_CLASS** table and the **CLASS** table.

The **ATT_CLASS** table will have a row for each unique combination of attendee and class. In other words, for each class that an attendee signs up for, there will be a row in the **ATT_CLASS** table that contains a copy of the primary key for the attendee and a copy of the primary key for the class.

The advantage in using a linking table is that it provides an unlimited many-to-many relationship between two tables, without any data duplication. It gives us the flexibility to assign as many classes to an attendee and as many attendees to a class as the application will allow.

From the **CLASS** table's point of view, the linking table indicates which attendees are enrolled in each class. From the **ATTENDEE** table's point of view, it indicates the list of classes in which a particular attendee has enrolled. In the next chapter, we will learn how to handle this many-to-many relationship with SQL and demonstrate some of these types of queries.

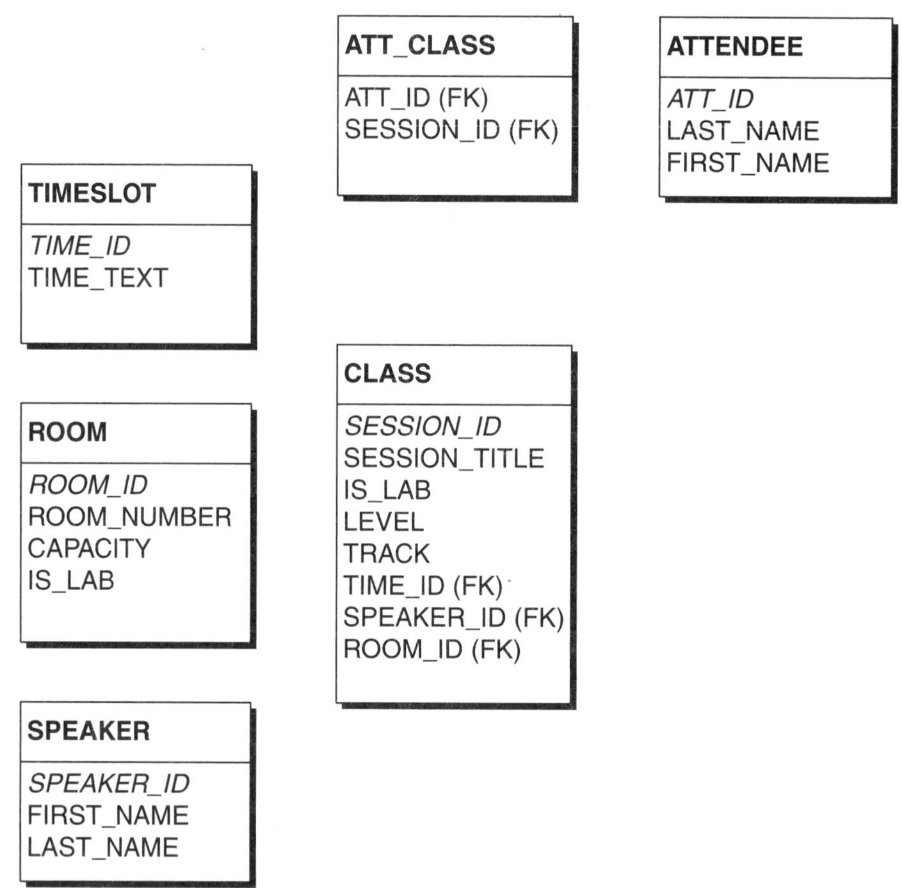

Figure 3.15 *Revised ER Diagram Showing Primary and Foreign Keys*

NORMALIZATION

Now that the primary entities have been established and the attributes have been assigned, it's time to look at our database from the viewpoint of efficiency. If you take the time to follow the process that has been outlined here, the normalization process becomes very easy. In some cases, if your entities are well defined, your design will usually pass first normal form without any modification. If all of your attributes have been assigned to the appropriate entities and there is no chance of duplication, you may even pass third normal form.

As you begin to understand the normal forms, and how they work, your design thought process will introduce those tests from the very beginning, and throughout the design development. Even so, it is always a good idea to re-check your database against the three normal forms before you call your design final.

Our conference database passes first normal form, because there are no repeating groups of attributes. Again, this is a result of having taken the time to interview the users and start with a solid set of entities. So often, it is tempting to begin with a single-table design, and work from there. The fact is that single-table designs simply don't work. Very few database applications can be successful with just a single table. This doesn't mean that your database *must* have more than one table. It simply means that you should look at the *needs* of the application first, then start developing a table structure—not the other way around.

Due to the simplicity of this example, the database is broken down into well-defined entities. With the exception of the **CLASS** table, the database complies with second normal form. In the **CLASS** table, however, we can see a couple of areas where duplication of data might occur.

The first change we may want to consider is to separate the information about levels (beginner, intermediate, and advanced, etc.) into its own table, and relate it back to the **CLASS** table with a foreign key. This is a case where you have a fixed number of options from which to choose for the value of an attribute. If we keep this field in the **CLASS** table, then the database can impose no restrictions on the values that are placed in that field. By listing the valid values in a separate table, you can gain more control over the user's ability to populate the field with correct data.

The second change that should be considered is to move the track (the class subject category) into a separate table. Similar to level, track is a field that has a fixed number of values that it should contain. For keeping the data consistent and accurate, a separate table is appropriate.

The result of applying second normal form is shown in Figure 3.16. As we look at each of the tables in this design, they each have a specific purpose, and the fields in each table help support that purpose. These are the fundamental characteristics of a good database design.

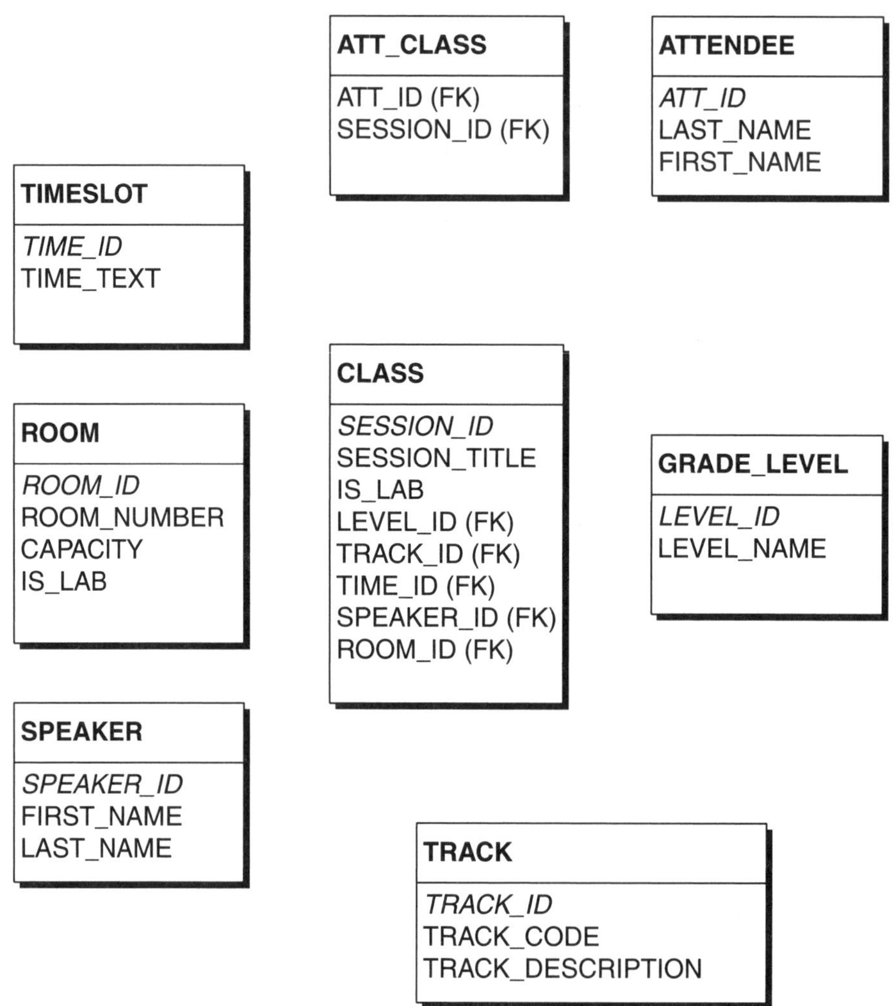

Figure 3.16 *Final ER Diagram for the Conference Database*

REFERENTIAL INTEGRITY

In the final stages of the design, we created the **GRADE_LEVEL** and **TRACK** tables to help eliminate redundant data and improve data accuracy. While these kinds of adjustments to the design are a fundamental part of the normalization process, they also accomplish another important goal of any database design: Let the database do as much of the data validation as possible.

The relationships in our ER diagram are not just for illustration purposes. They can be implemented in the relational database management system to help maintain data integrity. You can control how the database reacts to modifications made to tables that are a part of a relationship. *Referential integrity* is achieved when this type of control is implemented. For each individual foreign key/primary key relationship, you have three options:

- No restrictions
- Restrict
- Cascade

These options are discussed below

No Restrictions

In this scenario, you are saying that in a foreign key/primary key relationship, the database will allow a foreign key value that does not exist in the related table's primary key. In other words, you are **not** enforcing referential integrity in this particular relationship

This is rarely a desirable condition. It defeats the purpose of establishing the relationship in the first place. Without enforcing some sort of referential integrity, you risk having incorrect data, which makes your database difficult to maintain.

Restrict

When you restrict a relationship, the database requires that the foreign key value exist in the primary key of the related table. For add and update operations, this means that the database will reject any attempt to add or update a row that has an invalid foreign key value. For delete operations, the database will reject any attempt to delete a row that has a primary key referenced by a foreign key value in the related table.

In the conference database, you should implement at least this level of referential integrity on all relationships. For example, if a speaker cancels for some reason or another (a row in the **SPEAKER** table needs to be deleted), the database will require that no rows in the **CLASS** table exist that reference that speaker. This means that either the classes need to be deleted as well or they need to be assigned to another speaker.

Cascade

For add operations, this option behaves in the same way as the restrict option. The additional cascade option affects update and delete operations. For update, it means that the database allows primary key values to be updated, but all foreign key values that reference it are also updated in the related tables. For delete, it means that if a row is deleted that has a primary key referenced by one or more foreign key in other tables, the corresponding rows in the related tables will also be deleted. Since primary keys are rarely updated, this option is primarily associated with the delete operations.

A common place for the use of the cascade option is with linking tables in many-to-many relationships. For example, in the conference database, we would implement the cascade option for both relationships that include the **ATT_CLASS** table. With this type of restriction in place, when a row is deleted in the **ATTENDEE** table, all rows that referenced it in the **ATT_CLASS** table are no longer necessary and will also be deleted. Similarly, if a class is deleted, then references to it in the **ATT_CLASS** should also be deleted for each attendee that had enrolled in the class.

OTHER DATA INTEGRITY ISSUES

Through the methods described above, certain types of data integrity can be enforced. However, there will usually be other data integrity issues to deal with. Some of them can be enforced by the database, and others must rely on the application for proper enforcement.

The following are examples of conditions in the conference database that the database system cannot prevent:

- More than one session is assigned to a room during a single timeslot.
- An attendee is assigned to more than one class during a single timeslot.
- An attendee is assigned to the same class more than once.
- A speaker is assigned to more than one class during a single timeslot.
- A session is designated as a lab in a room that is not lab-equipped.
- The number of attendees enrolled in a class exceeds the capacity of the room.

If the database system cannot prevent these types of conditions, it then becomes the responsibility of the application to handle them. In the next chapter, we will show how SQL can be used to identify some of these conditions.

VALIDATING THE DESIGN

How do we know if our database design is a *good* database design? The merits of a database design can only be measured by how well it performs in the context of the application. The real reason we designed this database in the first place was to use it in an application, populate it with data, and get valuable information from it. In the next chapter, as we learn SQL, we will test the database design against several query scenarios. Testing the database's ability to provide the information we need through SQL is a great way to validate a database design.

During the design process, you should start thinking about the specific types of output your application will provide to its users. Again, the best way to understand this is though the interview process. Listed below are just a few examples of queries that may be asked of this database:

- Get a list of sessions, with the name of the speaker for each session.
- Determine the number of students enrolled for each session.
- List which speakers are teaching more than two classes.

SUMMARY

In this chapter, you have learned how to

- Understand the benefits of a well-designed database
- Work through the process of designing a database
- Develop an entity relationship diagram
- Identify primary and foreign keys
- Normalize a database
- Understand referential integrity

REVIEW QUESTIONS

1. What are some of the consequences of a poorly designed database?
2. What are two primary goals of database design?
3. What are the components of an entity relationship diagram?
4. What constitutes a good primary key?
5. What is the relationship between a primary key and a foreign key?
6. How is a many-to-many relationship handled?
7. What kinds of restrictions can a database impose on a relationship?

CHAPTER 4

Using SQL

OBJECTIVES

After completing this chapter, you will be able to

- Compose simple queries using **SELECT**
- Retrieve specific rows using a **WHERE** clause
- Build complex queries using multiple tables
- Understand the concept of the SQL **VIEW**
- Modify a database using **INSERT**, **UPDATE**, and **DELETE**
- Use SQL to check database integrity
- Understand the advantages and disadvantages of linking to a **VIEW**

INTRODUCTION

SQL is the standard language used to communicate with databases. With SQL, you have complete control of your database. You can add, change, or delete information in tables, and you can formulate queries that allow you to look at your data in different ways. To get the most out of your database, you need to have a basic understanding of what SQL can do for you and at least be able to develop some simple queries.

This chapter provides an overview of the capabilities of SQL. Rather than trying to be a comprehensive training manual on SQL, this chapter concentrates on its most commonly used features—specifically those features that are most useful within an AutoCAD/database application.

Most of our attention is focused on data manipulation and queries. Data manipulation refers to the adding, changing, and deleting of data. In SQL, the commands used for data manipulation are **INSERT**, **UPDATE**, and **DELETE**. Queries are the mechanism by which information is extracted from the database. In SQL, the command used to issue a query is **SELECT**.

The version of the SQL standard supported by AutoCAD 2000 is called "SQL2," also referred to as "SQL-92." Throughout the book, the term "SQL" is used to refer to this standard.

THE EXAMPLE CONFERENCE DATABASE

To illustrate some of these key concepts of SQL, the sample conference database you designed in the previous chapter is used. Figure 4.1 shows the final entity relationship diagram (ERD) that was developed.

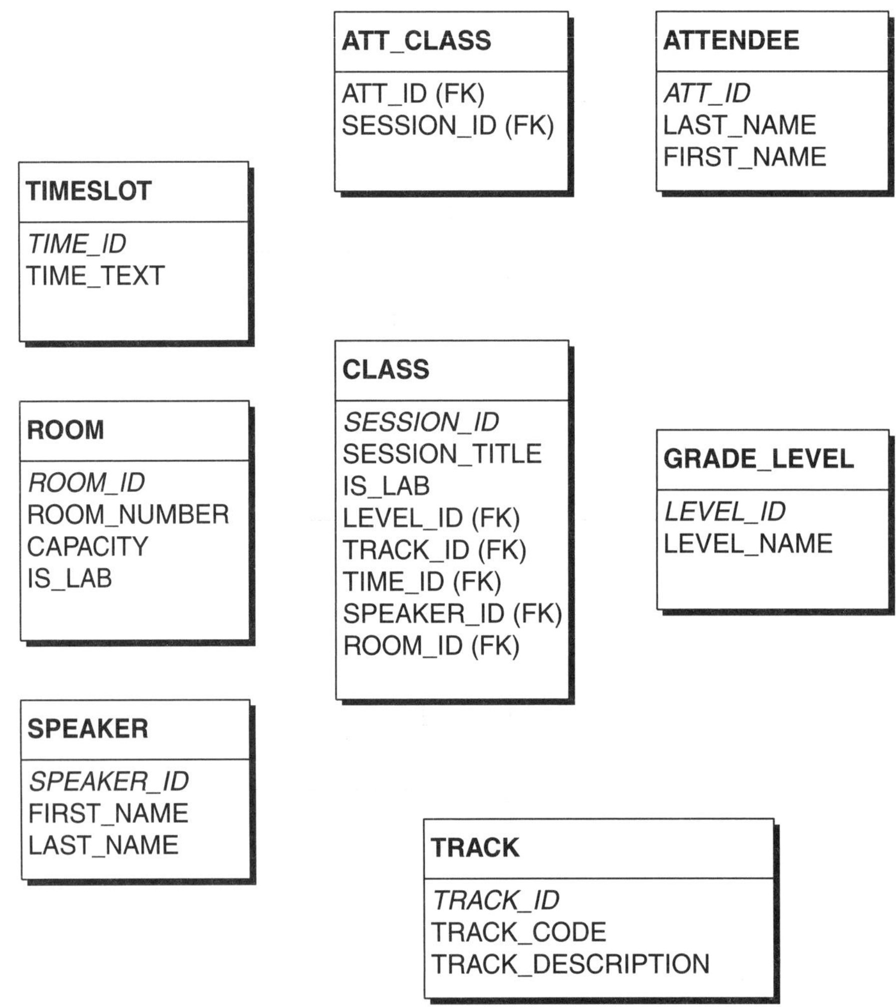

Figure 4.1 *The Conference Database*

EXECUTING THE SQL EXAMPLES

Naturally, this chapter includes several SQL statements that illustrate the various concepts and capabilities of SQL. All of the examples can be executed from within AutoCAD 2000 through the Query Editor. You will need to copy the class.mdb file from the CD-ROM.

There is also an exported "query set" file on the CD-ROM called class.dbq. This file contains all of the example queries in this chapter. Each query name in this file corresponds to the *table number* given to the output of each query throughout the chapter.

Before you can use the Query Editor to issue queries on the class database, you must create a data source. The following tutorial will take you through this process as well as show you how to import the query file.

TUTORIAL 4.1 – CONFIGURING A DATA SOURCE FOR THE CLASS DATABASE

1. Start AutoCAD 2000 with an empty drawing, and display the dbConnect Manager window.
2. Right-click the Data Sources branch in the dbConnect Manager and choose **Configure Data Source...** from the shortcut menu. The **Configure a Data Source** dialog box is displayed.
3. Type "class" as the data source name, and click **OK** or press ENTER. The **Data Link Properties** dialog box is displayed.
4. Select **Microsoft Jet 3.51 OLE DB Provider** in the list of OLE DB Providers, and click **Next >>**.
5. Enter the full path of the class.mdb file. This file is included on the CD-ROM.
6. Click **OK**.

Note: Steps 7 and 8 show you how to import the query set into the current drawing. If you would rather manually type the example queries, you can skip these steps.

7. From the **dbConnect** menu, choose **Queries**, and then **Import Query Set**. The **Import Query Set** dialog box is displayed.
8. Navigate to the location of the class.dbq file you copied from the CD-ROM, and choose **Open**. The queries appear as nodes under the drawing node.

THE BASIC RULES OF SQL

THE SQL HIERARCHY

SQL defines a hierarchy that includes environment, catalog, schema and table. Environments contain multiple catalogs, catalogs contain multiple schemas, and schemas contain multiple tables. Figure 4.2 illustrates the relationships between these components. You probably recognize this hierarchy from the discussion on link templates in Chapter 1.

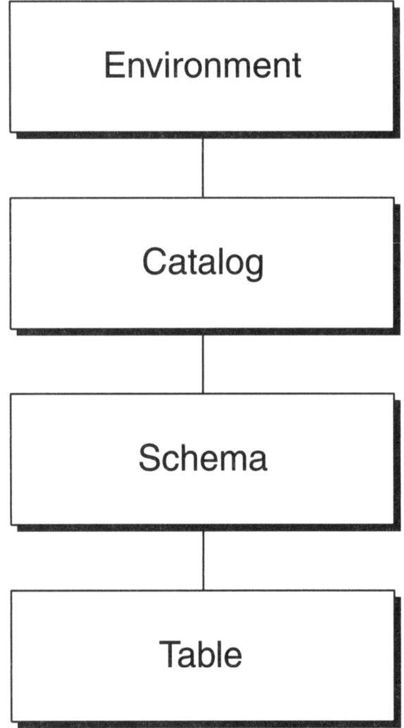

Figure 4.2 *The SQL Hierarchy*

Environment

The *environment* describes the framework for a particular system. It consists of a relational database management system (RDBMS), its associated databases, and information about users and programs and how they interact with the databases. In AutoCAD, the environment is also referred to as the *data source*. An environment typically contains one or more catalogs.

Catalog

A *catalog* represents a collection of related tables for a particular application. It is similar in concept to a *database* in previous versions of SQL. A catalog contains one or more schemas.

Schema

A *schema* is a collection of tables held within the catalog. In some database systems such as Oracle, the schema represents a single user's portion of the complete database. This means that when you log in to the database, you are logging in to a schema. A schema can contain one or more tables, and a table can be a member of more than one schema.

Table

A *table* is a two-dimensional array of *rows* and *columns*. If you picture a table as a rectangular grid, the rows run horizontally and the columns run vertically. A single row in a table generally contains a set of data that are related in some way. In fact, this is where the term *relational database* got its name. In some contexts, you may find that a *row* is sometimes referred to as a *record*, and *columns* are sometimes referred to as *fields*. Either way, they refer to the same concepts.

By definition, the rows of a table in SQL are *unordered*. This does not mean that you can't look at the data in some particular order. With SQL, you can look at the data sorted by any column you choose. For example, you might want to look at the **CLASS** table sorted in ascending order by session title. What provides this flexibility is the fact that the rows are in no particular order to begin with. The columns on the other hand, *are* ordered and numbered. So you can refer to the **session_title** column, which is the second column, as "2."

In order for a single row in a table to be identified, each row must be unique. Usually a single column is used for this purpose. In the **SPEAKER** table, for example, the numeric column **speaker_id** ensures that each row in the table is unique, even if two speakers have the same name. This column is referred to as the *primary key* of the table. From AutoCAD's perspective, if a table has a primary key, this column is typically used for linking purposes as well.

SQL DATA TYPES

SQL provides several data types that most database systems support. However, some database management systems provide additional data types that are not part of the SQL standard or they use different names to refer to similar SQL data types. For the purposes of this discussion of SQL, we will briefly discuss three of the most common data types found in any database, *character*, *numeric*, and *date*.

Character

The *character* data type is used to store strings containing any type of displayable character, including alphabetic characters, numeric characters and symbols. A column in a table that defines character data usually has a *size* associated with it, which controls the maximum number of characters that column allows in any particular row.

Character data can be sorted alphabetically but cannot be used in mathematical expressions such as **SUM**, which is described in more detail later. When you write SQL statements, character values are always delimited with *single quotes*.

Numeric

The *numeric* data type is used exclusively to store numeric values. SQL supports a variety of numeric formats, including integers, and decimal values. Integer values can be defined in various sizes, depending on the highest number you need to represent. Decimal values can be defined with a specific precision, or number of decimal places. Numeric values can be sorted, as well as used in mathematical expressions such as **SUM** or **AVG**, which are described in more detail later in this chapter.

Date

The *date* data type, or date/time data type, is a special type used to store a specific date and time. The format of the date value is important, and it usually consists of the day, month, and year separated by either a dash ("-") character or a slash ("/") character. In some systems, the order is reversed. For example, July 4th 1776 is stored as '17760704'. Date fields can be sorted, and most systems allow certain mathematical operations to be performed on them, such as addition and subtraction.

TYPES OF SQL COMMANDS

Generally speaking, SQL commands can be broken down into the following three categories:

Data Definition Language (DDL)—Used to create or remove entire objects from the database, such as tables or views. The SQL commands that are a part of DDL include

- **CREATE**
- **DROP**

Data Manipulation Language (DML)—Used to query and modify data within the database. The SQL commands that are a part of DML include

- **SELECT**
- **INSERT**
- **UPDATE**
- **DELETE**

Data Control Language (DCL)—Used to define user rights to perform various types of operations on a database. The SQL commands that are a part of DCL include

- **GRANT**
- **REVOKE**

This chapter focuses primarily on Data Manipulation Language. The majority of the SQL usage we, as AutoCAD users, encounter falls under this category.

USING SQL TO CREATE QUERIES

THE SELECT STATEMENT

One of the most common operations in SQL is the *query*. In fact, the origin of the acronym "SQL" comes from "Structured Query Language." The SQL keyword used to issue a query is **SELECT**. A simple example of a **SELECT** statement is to retrieve information from a single table.

```
SELECT speaker_id, first_name, last_name
FROM speaker
```

The output for this query is as follows:

speaker_id	first_name	last_name
1	Rebecca	Davis
2	Barry	Buckman
3	Todd	Crawford
4	Tim	Patterson

Table 4.1 *A Simple SELECT Statement*

This example shows how the **SELECT** command could be used to retrieve the entire contents of a table. Let's take a look at the syntax of this statement.

SELECT <*list of column names*> **FROM** <*table name*>

All **SELECT** statements must use at least the **SELECT** keyword and the **FROM** keyword, and the column names are always separated with commas. If you want to display all of the columns, you can use an asterisk (*) in place of the list of column names as follows:

```
SELECT * FROM speaker
```

This gives you the same result as the previous example.

The ability to provide a specific list of columns in the **SELECT** statement allows you to display only the columns you want to see, and in the order you want to see them. For example:

```
SELECT last_name, first_name
FROM speaker
```

This query produces the following output:

last_name	first_name
Davis	Rebecca
Buckman	Barry
Crawford	Todd
Patterson	Tim

Table 4.2 *A SELECT Statement with Reordered Columns*

THE WHERE CLAUSE

While the above examples demonstrate the retrieval of all rows in the table, **SELECT** is most often used to retrieve a *subset* of the rows. To accomplish this, you use the **WHERE** keyword along with a condition that must be met in order for a row to be returned. For example, if you wanted a list of the sessions that were rated at level 3, you could use the following SQL statement:

```
SELECT session_title
FROM class
WHERE level_id = 3
```

session_title
Modeling in 3D Studio Max
Using AutoLISP

Table 4.3 *A Simple WHERE Clause*

CONDITIONAL EXPRESSIONS

Following the **WHERE** keyword is a *conditional expression*, which is applied to each row in the table. What you're saying is, "show me only the rows for which the following condition is met." In this case, we used the 'equals' operator. SQL recognizes the same types of relational operators as other computer languages do. These are

=	Equal to
>	Greater than
<	Less than
>=	Greater than or equal to
<=	Less than or equal to
<>	Not equal to

So obtaining a list of the sessions that were rated at level 2 or greater is accomplished through the following SQL statement:

```
SELECT session_title, level_id
FROM class
WHERE level_id >= 2
```

session_title	level_id
Electromechanical Design	2
Customizing menus	2
Publishing Your Maps with MapGuide	2
Modeling in 3D Studio Max	3
Using AutoLISP	3

Table 4.4 *Using a Conditional Expression*

Note: Columns used in the **WHERE** clause do not necessarily need to be included in the output of the **SELECT** statement.

Boolean Operators: AND, OR, and NOT

You can also combine multiple expressions with the Boolean operators **AND** and **OR**. For example, consider the following query.

```
SELECT session_title, level_id
FROM class
WHERE level_id > 1
AND speaker_id = 2
```

session_title	level_id
Electromechanical Design	2
Using AutoLISP	3
Publishing Your Maps with MapGuide	2

Table 4.5 *A WHERE Clause using the AND Operator*

Each row is tested for two conditions: **class_level > 1** and **speaker_id = 1**. Since the Boolean operator **AND** was used, *both* expressions must be true for each row returned.

SQL also supports **OR**, which evaluates two expressions and if *either* expression is true, then the entire expression evaluates to true. The keyword **NOT** can be used to reverse the evaluation of any single Boolean expression. Here's a slightly more complex query demonstrating all three Boolean operators. This query also demonstrates how parentheses can be used to group expressions together when multiple operators are used.

Note: Some database systems, such as Microsoft Access, provide a Boolean data type. This data type is used in the **CLASS** table for the **is_lab** column. A column that is a Boolean data type can normally be used by itself in a **WHERE** clause as a Boolean expression. However, the Query Editor in AutoCAD may force you to use a comparison operator anyway. In this case, when **is_lab** is "True," its value is -1. When is_lab is "False," its value is 0.

```
SELECT session_title, level_id
FROM class
WHERE (level_id = 1 OR level_id = 2)
AND NOT is_lab = -1
```

session_title	level_id
Electromechanical Design	2
How to Use Grips	1
Customizing menus	2
Teaching AutoCAD	1
Increase Productivity using Paperspace	1
AutoCAD Tips and Tricks	1

Table 4.6 *Using Multiple Boolean Operators*

Using the LIKE Operator and Wild Cards

When the 'equals' operator is used to compare strings, it is a *case-sensitive* match. The **LIKE** operator can be used in place of an 'equals' operator when you do *case-insensitive* string comparisons. Strings compared with the **LIKE** operator can also contain *wild cards*. Wild cards are special characters used within the literal string you are using to make a comparison. Table 4.7 lists the wild-card characters supported by most systems and shows examples of each.

Wild-card Character	Representation	Examples	Matches
_	Any single character	'CA_'	'CAD', 'CAR', 'CAB'
%	Zero or more characters	'%Map%'	'AutoCAD Map', 'MapGuide'
[charlist]	Any single character in charlist	'[HDL]ewey'	'Hewey', 'Dewey', 'Lewey'
[!char list]	Any single character not in charlist	'McFarlan[!d]'	'McFarlane'
[char range]	Any single character in char range	'[M-V]an'	'Man', 'Pan', 'Ran', 'Tan', 'Van'
[!char range]	Any single character not in char range	'[!M-V]an'	'Ban', 'Can', 'Dan'

Table 4.7 *Wild Cards*

Note: Some database systems support additional wild-card characters or alternative characters for the representations described above. Check the documentation for your specific database system. For example, Microsoft Access uses '?' and '*' instead of '_' and '%'. However, if you are working with an Access database from within AutoCAD, you should still use '_' and '%'.

Let's look at a simple example of using the **LIKE** operator with wild cards. Suppose you want to find all the classes that have the word "map" in their title.

```
SELECT session_title
FROM class
WHERE session_title LIKE '%map%'
```

session_title
Publishing Your Maps with MapGuide
Introduction to AutoCAD Map

Table 4.8 *The LIKE Operator and Wild Cards*

Note: Notice that the query returned rows with the word 'Map' with an uppercase 'M', even though the query used lowercase.

SORTING OUTPUT WITH ORDER BY

You can change the order of the output of a query by sorting the values in a particular column. To do this, you use the SQL **ORDER BY** keyword, followed by one or more column names. You can also specify ascending or descending by adding either the **ASC** or **DESC** keyword after the column names. Ascending is the default. Let's look at the **CLASS** table again, listing the session titles sorted in descending order.

```
SELECT session_title
FROM class
ORDER BY session_title DESC
```

session_title
Using AutoLISP
Teaching AutoCAD
Publishing Your Maps with MapGuide
Modeling in 3D Studio Max
Introduction to AutoCAD Map
Increase Productivity using Paperspace
How to Use Grips
Electromechanical Design
Customizing menus
AutoCAD Tips and Tricks

Table 4.9 *Sorting Output with ORDER BY*

QUERYING FROM MULTIPLE TABLES

The power behind relational databases is the ability to establish relationships between tables, and in turn, to formulate queries that work with these relationships. Our example database consists of eight tables that are all related in some way. To get any meaningful information from this database, our queries almost certainly need to include multiple tables.

For example, our database defines a relationship between the **CLASS** table and the **SPEAKER** table. The **SPEAKER** table has a primary key called **speaker_id**, and the **CLASS** table has a foreign key called **speaker_id** that is used to establish the relationship. SQL allows you to *join* multiple tables into a single query by listing the tables immediately following the **FROM** keyword. The **WHERE** clause can then be used to match the primary key in the **SPEAKER** table with the foreign key in the **CLASS** table. The following query lists each class, with its session title and the first and last name of the speaker.

```
SELECT session_title, first_name, last_name
FROM class, speaker
WHERE class.speaker_id = speaker.speaker_id
```

session_title	first_name	last_name
Teaching AutoCAD	Rebecca	Davis
Increase Productivity using Paperspace	Rebecca	Davis
Electromechanical Design	Barry	Buckman
Using AutoLISP	Barry	Buckman
Publishing Your Maps with MapGuide	Barry	Buckman
How to Use Grips	Todd	Crawford
Customizing menus	Todd	Crawford
Modeling in 3D Studio Max	Tim	Patterson
AutoCAD Tips and Tricks	Tim	Patterson
Introduction to AutoCAD Map	Tim	Patterson

Table 4.10 *Querying from Multiple Tables*

When multiple tables are specified after the **FROM** keyword, SQL returns every possible combination of rows from all tables. In order for the rows to be displayed correctly according to the relationships, the relationship must be specified in the query. In the example above, we are using the **WHERE** clause to establish the relationship between the **CLASS** table and the **SPEAKER** table using the **speaker_id** column in each table.

Note: Since both tables contain a **speaker_id** column, the column name is prefixed with the table name followed by a period. This syntax must be used any time a query references column names that are not unique across the tables being queried.

Here, we are gathering information from five separate tables in our database.

```
SELECT track_code, session_title, last_name,↵
room_number, time_text
```

```
FROM class, room, speaker, timeslot, track
WHERE class.room_id = room.room_id
        AND class.speaker_id = speaker.speaker_id
        AND class.time_id = timeslot.time_id
        AND class.track_id = track.track_id
```

track_code	session_title	last_name	room_number	time_text
VIS	Modeling in 3D Studio Max	Patterson	101A	10:30 a.m. - 12:00 p.m.
MEC	Electromechanical Design	Buckman	101B	10:30 a.m. - 12:00 p.m.
CAD	How to Use Grips	Crawford	102	10:30 a.m. - 12:00 p.m.
PRO	Using AutoLISP	Buckman	101A	1 p.m. - 2:30 p.m.
PRO	Customizing menus	Crawford	101B	1 p.m. - 2:30 p.m.
EDU	Teaching AutoCAD	Davis	102	1 p.m. - 2:30 p.m.
GIS	Publishing Your Maps with MapGuide	Buckman	101A	3 p.m. - 4:30 p.m.
CAD	Increase Productivity using Paperspace	Davis	101B	3 p.m. - 4:30 p.m.
CAD	AutoCAD Tips and Tricks	Patterson	102	3 p.m. - 4:30 p.m.
GIS	Introduction to AutoCAD Map	Patterson	103	1 p.m. - 2:30 p.m.

Table 4.11 *Querying from Multiple Tables*

AGGREGATE FUNCTIONS

Once you understand how to query across multiple tables, you can start gathering some valuable information from your database. A typical requirement for a database application is to summarize the information in the database. In SQL, this is accomplished through *aggregate functions*. The aggregate functions supported by SQL are

COUNT—returns the number of rows that the query selected

SUM—produces the mathematical sum of the selected numeric values

AVG—calculates the average of the selected numeric values

MAX—returns the largest of the selected values

MIN—returns the smallest of the selected values

Aggregate functions are applied to specific columns immediately following the **SELECT** keyword. **SUM** and **AVG** accept only numeric arguments, while **COUNT**, **MAX**, and **MIN** accept both numeric and character arguments. Let's look at some examples.

To find the total capacity of all the rooms in the **ROOM** table, you could use the following SQL statement:

```
SELECT SUM(capacity)
FROM room
```

364

Table 4.12 *Using the SUM Aggregate Function*

To find the total number of attendees at the conference (in other words, the number of rows in the **ATTENDEE** table)

```
SELECT COUNT(*)
FROM attendee
```

107

Table 4.13 *Using the COUNT Aggregate Function*

Creating Grouped Output with GROUP BY

The aggregate function examples shown above give you the result of the aggregate function applied to a single column on *all* of the rows selected. In other words, they always produce a single-column, single-row result. In some cases you may want to group rows together based on a particular value and apply the aggregate function to each group. For example, suppose you want to know the number of students enrolled in each session. This can be accomplished with a query that combines the **ATT_CLASS** table with the **CLASS** table as follows:

```
SELECT session_title, COUNT(att_id)
```

```
FROM att_class, class
WHERE att_class.session_id = class.session_id
GROUP BY session_title
```

session_title	
AutoCAD Tips and Tricks	56
Customizing menus	40
Electromechanical Design	39
How to Use Grips	33
Increase Productivity using Paperspace	21
Modeling in 3D Studio Max	36
Publishing Your Maps with MapGuide	32
Teaching AutoCAD	15
Using AutoLISP	50

Table 4.14 *Grouping Output with GROUP BY*

Using the HAVING Keyword

When you use the **GROUP BY** clause, it is sometimes necessary to apply a conditional expression to each of the grouped rows, rather than each of the original rows. For example, suppose you want to know which speakers are teaching more than two classes:

```
SELECT last_name, COUNT(session_number)
FROM speaker, class
WHERE class.speaker_id = speaker.speaker_id
  GROUP BY last_name
  HAVING COUNT(session_number) > 2
```

last_name	
Buckman	3
Patterson	3

Table 4.15 *Using the HAVING Keyword*

When you use both a **WHERE** clause and a **HAVING** clause, it is important to understand the order in which these conditional expressions are applied. The **WHERE** clause is always applied first. In this example, the **WHERE** clause establishes the relationship between the two tables using the **speaker_id** column. Next, the grouping takes place, and the aggregate function **COUNT** is applied. Then each group that is returned is tested against the **HAVING** clause.

Similar to the **WHERE** clause, columns used in the **HAVING** expression do not necessarily need to be included in the **SELECT** output. For example, if you were not concerned about showing the actual number of sessions each speaker was teaching in the previous query (all you wanted was the last name), you could write the query as follows:

```
SELECT last_name
FROM speaker, class
WHERE class.speaker_id = speaker.speaker_id
  GROUP BY last_name
  HAVING COUNT(session_number) > 2
```

Changing the Output Columns

SQL provides a way to change the name of the columns that are output from the query. This is accomplished with the **AS** keyword immediately following the column name you wish to rename. The syntax is as follows:

<old name> **AS** *<new name>*

This capability is especially useful when aggregate functions are being used. You probably noticed in the examples above that the aggregate functions return an undefined column name. Some database systems provide an arbitrary name for you, such as "**expr001**," but there is no way to tell exactly what the name will be.

For example, the query in the previous example could be rewritten as

```
SELECT last_name, COUNT(session_number) AS↵
num_classes
FROM speaker, class
WHERE class.speaker_id = speaker.speaker_id
  GROUP BY last_name
  HAVING COUNT(session_number) > 2
```

last_name	num_classes
Buckman	3
Patterson	3

Table 4.16 *Renaming the Output Column*

Note: Using **AS** does not physically change the name of the column in the database. It merely uses the new name for the query result.

ADVANCED QUERY CONCEPTS

USING VIEWS

As you work with a normalized database like this conference example, you may find that your SQL statements begin to get quite complex. Nearly every query that provides any meaningful information requires the use of aggregate functions and multiple tables. We can greatly simplify the query process with a concept called a *view*.

A view is basically a description of a *virtual table*. A virtual table is defined through a query. Views are created with the SQL **CREATE VIEW** statement. Not all database systems support views (such as FoxPro® 2). But if you are using a system that does support them, views can prove to be a very powerful and useful capability.

In general, operations can be performed on views just as they can be performed on normal tables. Queries can be performed on views simply with the inclusion of the view name in the **FROM** part of a **SELECT** statement. More importantly, AutoCAD objects can be linked directly to views, just as they can to tables. This technique is demonstrated later in this chapter.

Nesting Aggregate Functions

In our example conference database, the student enrollment in each class is something that must be continuously monitored and analyzed. It would be extremely useful if this number were simply another column in the **CLASS** table. For example, suppose you want to know the average number of students enrolled in all the classes. Unfortunately this is not possible with a single SQL query, because it requires the **AVG** aggregate function to be applied to the output of an already aggregated query through **COUNT**.

We can, however, accomplish this quite easily using a view. To do this, we will perform the following steps:

1. Compose a query that returns the enrollment for each class.
2. Store the query as a view called **ENROLLMENT**.
3. Use the **ENROLLMENT** view in a query that returns the average enrollment in all the classes.

Step 1: Composing the Query

The **ATT_CLASS** table supports the many-to-many relationship between the **CLASS** table and **ATTENDEE** table. Its only purpose is to track which attendees (students) are enrolled in each class (session). It contains just two columns; **att_id**, which defines the relationship with the **ATTENDEE** table, and **session_id**, which defines the relationship with the **CLASS** table.

Using a simple query with this table, we can obtain the total number of students enrolled, grouped by the session ID:

```
SELECT session_id, COUNT(att_id) AS enrolled
FROM att_class
GROUP BY session_id
```

This produces the output shown in Table 4.17.

session_id	enrolled
1	36
2	39
3	33
4	50
5	40
6	15
7	32
8	21
9	55

Table 4.17 *Enrollment by Session*

Step 2: Storing the Query as a View

The result of the previous query is a virtual table that has one row for each class. Each row contains the session ID, along with the count of attendees. You can give this virtual table a name and store its definition in the database with the SQL **CREATE VIEW** command:

```
CREATE VIEW enrollment AS
SELECT session_id, COUNT(att_id) AS enrolled
  FROM att_class
  GROUP BY session_id
```

Note: If you are following along with the class database from the CD-ROM, it is not necessary for you to execute this SQL statement. This **VIEW** is already defined in the database, and attempting to recreate it will result in an error.

CREATE VIEW is considered to be a part of Data Definition Language (DDL). Because this SQL statement does not start with **SELECT**, no rows are actually returned.

Step 3: Using the View in a Query

Once the view has been defined, it can be used in a **SELECT** statement just like any other table. You can now get the average number of students enrolled using the view name in the **FROM** clause:

```
SELECT AVG(enrolled)
FROM enrollment
```

35.77777777

Table 4.18 *Average Enrollment*

Creating a VIEW of the CLASS table

The **CLASS** table is really the focal point of the entire database. It therefore will be helpful if we have a single view of the **CLASS** table that includes the pertinent information from all related tables. We can also include our **ENROLLMENT** view as part of the query. Here's how this view can be created:

```
CREATE VIEW class_view AS
SELECT class.session_id, track_code,↵
track_description, session_title,↵
level_name, enrolled, last_name,↵
first_name, room_number, capacity,↵
room.is_lab, time_text, class.is_lab
  FROM class, room, grade_level, speaker,↵
timeslot, track, enrollment
      WHERE class.room_id = room.room_id
      AND class.level_id =  grade_level.level_id
      AND class.speaker_id = speaker.speak-↵
er_id
      AND class.time_id = timeslot.time_id
      AND class.track_id = track.track_id
      AND class.session_id =  enrollment.session_id
```

Note: If you are following along with the class database from the CD-ROM, it is not necessary for you to execute this SQL statement. This **VIEW** is already defined in the database, and attempting to recreate it will result in an error.

Using this view will greatly simplify queries on the **CLASS** table that require information from the other tables. Instead of using the ID columns to represent rows in the other tables, you can more easily use the columns that are more meaningful.

To see the entire view is as simple as

```
SELECT * FROM class_view
```

Also, you can now perform more complex queries on class enrollment with the new enrolled column. To illustrate this, **CLASS_VIEW** is used in some of the examples that follow.

QUERYING A TABLE AGAINST ITSELF

Occasionally, you may want to compare information in a table against other information within the same table. To distinguish the fields in one instance of the table from fields in the second instance, you must create *aliases* for the tables. The **AS** keyword, described earlier, can be used for this purpose.

For example, you could find all the sessions whose enrollment is at least as much as that of any session presented by the speaker whose last name is Patterson:

```
SELECT DISTINCT A.session_title, A.last_name,↵
A.enrolled
  FROM class_view AS A, class_view AS B
  WHERE A.enrolled >= B.enrolled
    AND B.last_name='Patterson'
    ORDER BY 3
```

session_title	last_name	enrolled
Modeling in 3D Studio Max	Patterson	36
Electromechanical Design	Buckman	39
Customizing menus	Crawford	40
Using AutoLISP	Buckman	50
AutoCAD Tips and Tricks	Patterson	55

Table 4.19 *Querying a Table Against Itself*

If you want to exclude the sessions given by Patterson from the output, you simply add the appropriate expression to the **WHERE** clause.

```
SELECT DISTINCT A.session_title, A.last_name,↵
A.enrolled
```

```
FROM class_view AS A, class_view AS B
WHERE A.enrolled >= B.enrolled
  AND B.last_name='Patterson'
  AND A.last_name <> 'Patterson'
  ORDER BY A.enrolled
```

session_title	last_name	enrolled
Electromechanical Design	Buckman	39
Customizing menus	Crawford	40
Using AutoLISP	Buckman	50

Table 4.20 *Querying a Table Against Itself*

In the previous two examples, you may have noticed the use of the **DISTINCT** keyword, which has not been discussed up to this point. The **DISTINCT** keyword is used to eliminate duplicate rows that exist in the results of a query. Remember, from our earlier discussion on querying multiple tables, that SQL tests the criteria in the **WHERE** clause to every possible combination of rows from all tables included in the **FROM** clause. Consider the following SQL statement:

```
SELECT A.session_title,A.enrolled
  FROM class_view AS A, class_view AS B
```

If there are nine rows in **CLASS_VIEW**, this query returns 9 x 9, or 81 rows. Each row is compared with itself and every other row in the table. Now add the next part of our query to compare the enrollment values:

```
SELECT A.session_title,A.enrolled
  FROM class_view AS A, class_view AS B
  WHERE A.enrolled > B.enrolled
```

This query still returns many duplicate rows. The number of duplicate rows returned for each session is based on the number of *other* sessions its enrollment is greater than. In fact, the only session that doesn't appear in the output is the session that has the lowest enrollment. If we add the **DISTINCT** keyword to the query above, all duplicate rows are eliminated and exactly one less than the total number of rows in the **CLASS** table is returned, or eight rows.

EMBEDDING A QUERY INSIDE ANOTHER QUERY

Another way to approach the previous example is to first find the minimum number of people enrolled in any class taught by Patterson using the **MIN** aggregate function:

```
SELECT MIN(enrolled)
  FROM class_view
  WHERE last_name = 'Patterson'
```

36

Table 4.21 *Using the MIN Aggregate Function*

This query returns a single value. In SQL, a query can return a value, or even a set of values, that can be used as part of an expression in another query. A query that is embedded inside another query is generally referred to as a *subquery* or a *nested query*. The value returned in the example above can be treated like any other number in a conditional expression. So we can embed this query inside another query to accomplish the original task in a single query:

```
SELECT session_title, last_name, enrolled
  FROM class_view
  WHERE enrolled >= (SELECT MIN(enrolled)
        FROM class_view
        WHERE last_name = 'Patterson')
    ORDER BY enrolled
```

This query produces exactly the same result as shown in Table 4.19.

SOME and ALL

When you embed a query that returns more than one row inside another query, you can use either the **SOME** or the **ALL** keyword in conjunction with a comparison operator. When **SOME** is used, the condition must be met for at least one of the values returned by the subquery. When **ALL** is used, the condition must be met for all of the values returned by the subquery.

Taking out the **MIN** function and adding the use of the **SOME** keyword, we have yet another way to write the same query:

```
SELECT session_title, last_name, enrolled
  FROM class_view
```

```
WHERE enrolled >= SOME (SELECT enrolled
          FROM class_view
          WHERE last_name =  'Patterson')
         ORDER BY enrolled
```

Again, this query produces exactly the same result as shown in Table 4.19.

Using the IN Keyword

Probably the most common use of subqueries is to determine if column values in one query exist in the output of another. This is accomplished with the **IN** keyword. Shown below is an example query that finds all the rooms that have a class assigned to them.

```
SELECT * FROM room
  WHERE room_id IN (SELECT room_id FROM  class)
```

You can also use the **NOT** operator in conjunction with **IN** to get the opposite effect. Suppose you want to get a list of rooms that do not have a class assigned to them:

```
SELECT * FROM room
  WHERE room_id NOT IN (SELECT room_id FROM  class)
```

USING SQL TO MODIFY THE DATABASE

THE INSERT COMMAND

Adding Individual Rows to Tables

The next few examples illustrate how you might register a new attendee named Jeff Lucas and sign him up for some classes. This process involves adding a new row to the **ATTENDEE** table and adding a row to the **ATT_CLASS** table for each class in which he will be registered.

To add new rows to a table, you use the **INSERT** keyword along with a list of column values for the new row to be added. For example:

```
INSERT INTO attendee
VALUES (114, 'Lucas', 'Jeff')
```

In the example above, values for every column in the table are included and they are in the correct order. You can also specify the column names, in any order, enclosed in parentheses immediately following the table name. The values then must also be specified in the same order.

```
INSERT INTO attendee (first_name, last_name,↵
att_id)
VALUES ('Jeff','Lucas',114)
```

Adding Multiple Rows to Tables

The **INSERT** command can also be used to append the output of a **SELECT** statement to a table. The syntax for this is as follows:

INSERT INTO *<table name>*
SELECT *<list of column names>*
FROM *<table name>*
WHERE *<expression>*

The **SELECT** statement used can be any valid **SELECT** statement. The only requirement is that columns that are returned by the **SELECT** statement must be valid columns in the table to which you are appending the new rows.

So the next step in our example is to add rows to the **ATT_CLASS** table. The **ATT_CLASS** table has two columns, **att_id** and **session_id**, both of which must have valid values. All the new rows for this attendee will have a value of 114 in the **att_id** column. For the purposes of this example, suppose Mr. Lucas wants to attend exactly the same classes as another attendee named Bob Myer.

You can get a list of the sessions that Bob Myer is attending with the following **SELECT** statement:

```
SELECT session_id
  FROM attendee, att_class
  WHERE attendee.att_id = att_class.att_id
    AND first_name = 'Bob'
    AND last_name = 'Myer'
```

SQL permits the use of constant values in the output of a **SELECT** statement, so we can use the constant value 114 to represent the **att_id** column as follows:

```
SELECT 114 AS att_id, session_id
  FROM attendee, att_class
  WHERE attendee.att_id = att_class.att_id
    AND first_name = 'Bob'
    AND last_name = 'Myer'
```

This produces the result shown in Table 4.22.

att_id	session_id
114	1
114	5
114	9

Table 4.22 *SELECT Using a Constant Value*

Now that we have exactly what is to be appended to the **ATT_CLASS** table, we can use this **SELECT** statement in an **INSERT** command to add the new rows:

```
INSERT INTO att_class
SELECT 114 AS att_id, session_id
  FROM attendee, att_class
  WHERE attendee.att_id = att_class.att_id
    AND first_name = 'Bob'
    AND last_name = 'Myer'
```

THE UPDATE COMMAND

The **UPDATE** command is used to modify information in a table. The syntax is as follows:

> **UPDATE** <*table name*>
> **SET** <*column name*>=<*value*>, [<*column name*>=<*value*>], ...
> **WHERE** <*expression*>

Note: The **WHERE** clause in the **UPDATE** command is optional. If no **WHERE** clause is used, the **UPDATE** affects all rows in the table.

Suppose you need to relocate the session titled 'Using AutoLISP' from room '101A' to room '102.' This operation involves changing the **room_id** value for this session to 3, the **room_id** for room 102. The **UPDATE** command is written as follows:

```
UPDATE class
SET room_id=3
WHERE session_title='Using AutoLISP'
```

SQL also permits the use of certain expressions in the **SET** portion of the **UPDATE** command. For example, suppose you decide that you can increase the capacity of all the rooms in your conference center by ten percent. This can be accomplished with the following statement:

```
UPDATE room
SET capacity = capacity * 1.1
```

THE DELETE COMMAND

The **DELETE** command is used to remove rows from a table. The syntax is as follows:

> **DELETE FROM** <*table name*>
> **WHERE** <*expression*>

Note: The **WHERE** clause in the **DELETE** command is optional. If no **WHERE** clause is used, all rows in the table are deleted.

Suppose you want to delete the new attendee that was added in an earlier example. If you know that the ID of the attendee is 114, then the **DELETE** statement is as follows:

```
DELETE FROM attendee
WHERE att_id = 114
```

In SQL, you have the ability to establish relationships between tables that ensure referential integrity is maintained. For example, in our example database, the relationship between the **ATTENDEE** table and the **ATT_CLASS** table stipulates that if an attendee is deleted, all rows in the **ATT_CLASS** table with that attendee ID are also deleted. In other words, this **DELETE** command not only deleted a row from the **ATTENDEE** table; it also deleted three rows from the **ATT_CLASS** table.

USING SQL QUERIES FOR DATA VALIDATION

An important part of implementing a database application is making sure the data is valid and makes sense in its reflection of reality. There are certain data integrity issues that the database itself can take care of, and others that your application must address. SQL can be an important tool in detecting and addressing such issues.

In the Class database, while much of the referential integrity is set up to be handled by the database, there is plenty of opportunity for the data to be incorrect. In Chapter 3, we identified the following conditions that the database could not prevent:

- More than one session is assigned to a room during a single timeslot.
- An attendee is assigned to more than one class during a single timeslot.
- An attendee is assigned to the same class more than once.
- A speaker is assigned to more than one class during a single timeslot.
- A session is designated as a lab in a room that is not lab-equipped.
- The number of attendees enrolled in a class exceeds the capacity of the room.

Let's take a look at how SQL can be used to identify some of these problems.

More than one session is assigned a room during a single timeslot.

Do you notice anything common among the first four problems listed above? They all have the words "more than one" in them. This should tell you that the SQL statements used to detect these problems will also be similar. If you are trying to find "more than one" of something in a table, the easiest way to accomplish this is by grouping the output of the query on the columns you want to display, then follow that with **HAVING COUNT(*) > 1**.

Simply detecting if more than one session has been assigned to a room during a single timeslot is quite simple:

```
SELECT room_id, time_id
  FROM class
GROUP by room_id, time_id
    HAVING COUNT(*) > 1
```

The problem with this is that it only tells us the **room_id** and **time_id** of any instances where this condition exists. It does not tell us the *names* of the sessions. You would think that simply adding the **session_title** to our output would solve this problem. Consider the following query:

```
SELECT session_title, room_id, time_id
  FROM class
  GROUP by session_title, room_id, time_id
    HAVING COUNT(*) > 1
```

Unfortunately, this does not produce the desired result. This query only detects a condition in which all three columns are duplicated. You are really looking for *different* session titles, but duplicate rooms and timeslots.

This makes the problem a bit more complex. One approach would be to query the **CLASS** table against itself to produce duplicate rows where the rooms and times were equal. Then group the output and check for duplicates using **HAVING COUNT(*) > 1**.

```
SELECT A.session_title, A.room_id, A.time_id
  FROM class A, class B
  WHERE A.room_id = B.room_id
    AND A.time_id = B.time_id
    GROUP BY A.session_title, A.room_id,↵
A.time_id
      HAVING COUNT(*) > 1
```

Another method is to use a subquery:

```
SELECT session_title, room_id, time_id
  FROM class A
  WHERE room_id IN
    (SELECT room_id
      FROM class B
      GROUP BY room_id, time_id
        HAVING COUNT(*) > 1
        AND A.time_id = B.time_id)
```

Both of these queries return the same result, which is a list of the sessions assigned to the same room during the same timeslot. If no rows are returned, then the database passes this validation.

An attendee is assigned to more than one class during a single timeslot.

In the previous validation example, you were looking for duplicates of two columns (**room_id** and **time_id**) in a single table (**CLASS**). The following example shows how you can find duplicate rows that exist in a query across multiple tables.

```
SELECT first_name, last_name, time_id,  COUNT(*)
        FROM class, att_class, attendee
```

```
WHERE att_class.session_id=class.session_id
  AND attendee.att_id=att_class.att_id
  GROUP BY first_name, last_name,time_id
    HAVING COUNT(*) > 1
```

An attendee is assigned to the same class more than once.

This is simply a check for duplicate rows in the **ATT_CLASS** table. This type of condition is easy to detect but difficult to correct, because SQL cannot distinguish between two rows that contain exactly the same data.

```
SELECT att_id, session_id, COUNT(*)
  FROM att_class
  GROUP BY att_id, session_id
    HAVING COUNT(*) > 1
```

A session is designated as a lab in a room that is not lab-equipped, or vice versa.

This validity check is actually quite simple, especially if you use the **CLASS_VIEW** view we created earlier.

```
SELECT class_view.session_title,↵
class_view.class.is_lab,  class_view.room.is_lab
  FROM class_view
  WHERE class_view.class.is_lab <>  room.is_lab
```

The remaining two examples are left for you to try on your own. You will find them in the Exercises section at the end of this chapter.

PUTTING SQL TO WORK IN AUTOCAD

So far, this example has demonstrated how SQL can be used to work with the database in a pure database environment. What benefits do you, as an AutoCAD user, gain from such knowledge? The answer is simple. Knowledge of SQL can help you turn a simple database linked to AutoCAD into a robust application. In the chapters that follow, we will focus specifically on creating real applications with the tools that AutoCAD provides. As you read through these chapters, you will notice how critical SQL is to building a complete application.

But before we complete our discussion on SQL, let's take our conference example to the next level and demonstrate exactly how we can bring this database to life within AutoCAD.

DECIDING WHERE TO MAKE THE LINK

Suppose you have a drawing of a conference center, and you want to use it to graphically display the results of queries on the database. The most logical table to link to in our database is the **ROOM** table, because a room is something that can be depicted on a floor plan with a single graphic object, such as a closed polyline. Finally, since our **ROOM** table is already a separate table, and it has a single column that is used to uniquely identify each row, this database is well suited for use with AutoCAD.

Figure 4.3 shows the Class database model with the room polyline relationship depicted.

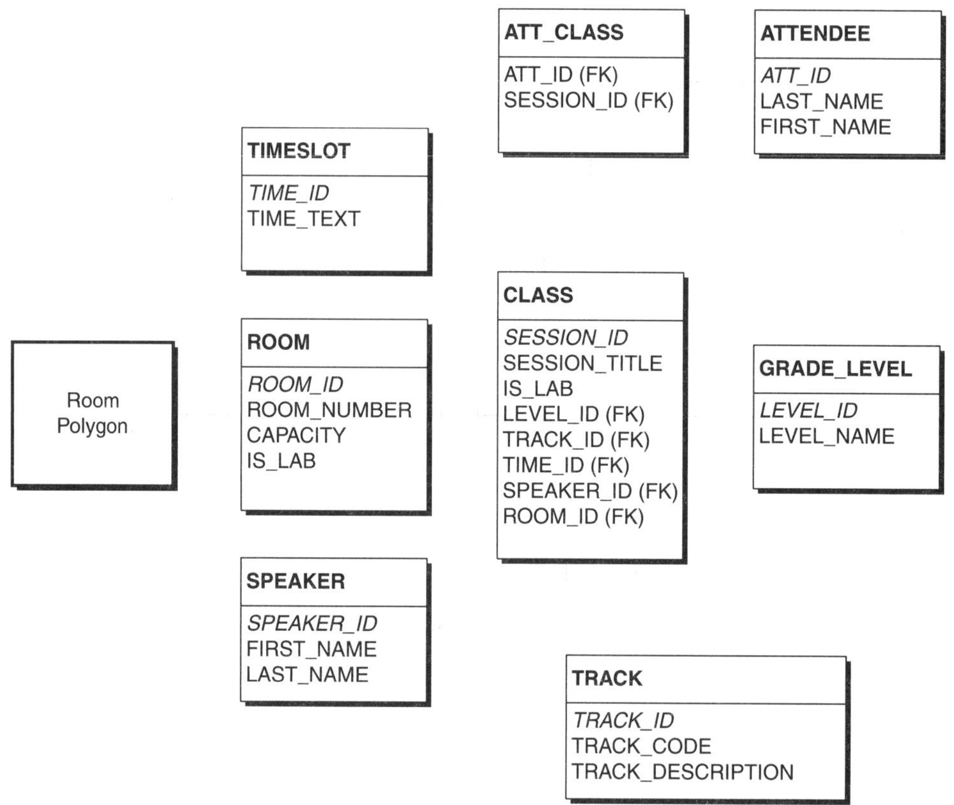

Figure 4.3 *Data Model with Link to AutoCAD*

LINKING A DRAWING TO THE CLASS DATABASE

The next step is to use dbConnect to link each room object to its corresponding row in the **ROOM** table. In our example, there are only five rooms that need to be represented, so creating a drawing with five objects and linking them is relatively easy. In a real application, in which there could potentially be hundreds of objects to link, you may want to consider automating, or partially automating, the linking process. In Chapter 7, you will learn exactly how this can be accomplished with VBA or AutoLISP.

The following tutorial takes you through the process of creating a simple drawing and linking it to the Class database.

TUTORIAL 4.2 – LINKING A DRAWING TO THE CLASS DATABASE

1. Start AutoCAD 2000 with an empty drawing.
2. Display the dbConnect Manager and connect to the class data source.
3. Right-click the ROOM table and choose **New Link Template**.
4. Keep the default name of **ROOMLink1** and click **Continue**.
5. Check the **room_id** column as the linking column, and click **OK**.
6. From the **Draw** menu, choose **Rectangle**.
7. Draw a small rectangle (for the purpose of this example, scale is not important).
8. Repeat Steps 6 and 7 until you have created five rectangles on the screen.
9. In the dbConnect Manager, right-click the ROOMLink1 link template and choose **Edit Table** from the shortcut menu.
10. Position or dock the Data View window so that the five rectangles are visible.
11. Highlight the first record and click the **Link!** toolbar button.
12. Select one of the rectangles and press ENTER.
13. Continue the process until each rectangle is linked to a unique row in the table.
14. Save the drawing, and call it room.dwg.

GRAPHICALLY SHOWING QUERY RESULTS

Now that the links have been established, you can start using the dbConnect interface to graphically select linked objects based on a query. In Chapters 1 and 2 you learned how to do this using the Query Editor. The examples that follow show how SQL can be used (in the Query Editor) to get a variety of information from the database. The results of the queries can be depicted with the linked room objects you just created.

Note: If you want to follow along using the Query Editor, use the **SQL Query** tab and make sure that both **Indicate records in data view** and **Indicate objects in drawing** are checked.

Because the AutoCAD objects are linked only to the **ROOM** table, this table must be included in any query we want to represent graphically. In common language, this means our questions must start with something like "Show me all the *rooms* that have…."

For example, the result of a question like, "Show me all the rooms that have a capacity greater than 50 could be represented graphically with the following SQL query:

```
SELECT * FROM room
  WHERE capacity > 50
```

Another simple example might be to find Room 101B.

```
SELECT * FROM room
  WHERE room_number = '101B'
```

Querying Multiple Tables

Querying information from the **ROOM** table alone is easy, but you will very quickly run out of questions to ask. To get anything more out of the database, you will probably want to gather some information from other tables in the database. This is where your knowledge of SQL will really start to pay off!

Suppose you want to find the room where the session titled 'Using AutoLISP' is being held. Normally, you could use the following query:

```
SELECT *
  FROM room, class
  WHERE class.room_id = room.room_id
    AND session_title = 'Using AutoLISP'
```

If you are following along, you probably noticed that while the results of this query appeared in the Data View window, no polylines were highlighted. Unfortunately, dbConnect's selection feature does not work with a query that uses more than one table in the **FROM** clause.

To ensure that the **Indicate objects in drawing** feature of the dbConnect Query Editor works properly, you must begin each query with

```
SELECT * FROM room WHERE ...
```

This makes querying information from other tables a bit tricky. There are, however, a number of workarounds. For example, you could link each room object, through the **room_id**, to the **CLASS** table for each class that takes place in the room. Then you could use the following query to get the desired result:

```
SELECT *
  FROM class
  WHERE session_title = 'Using AutoLISP'
```

The problem with this approach is that you are linking a table that represents a class to an AutoCAD object that represents a room. Keep in mind that each room has multiple classes. Getting them all linked requires having multiple links on each AutoCAD object, and this can quickly become a maintenance nightmare.

A more suitable approach would be to stick with the single link to the **ROOM** table and take advantage of SQL's ability to embed a query inside another query.

```
SELECT * FROM room
  WHERE room_id IN
  (SELECT room_id
    FROM class
    WHERE session_title = 'Using AutoLISP')
```

This query works with dbConnect's selection feature while allowing you to maintain a single link on each object.

HANDLING ONE-TO-MANY RELATIONSHIPS

The relationship between the **ROOM** table and its immediate neighbor, the **CLASS** table, is a one-to-many relationship. In other words, within each room there are multiple classes. Also, each class has multiple attendees. If you apply what you have learned about using SQL to query across multiple tables, these relationships should not be intimidating to you. As long as your query displays information in terms of the **ROOM** table (since that is the table we are linking to), there is no limit to the number of other tables that can be included in the query.

The previous example—to find the room where the session titled 'Using AutoLISP' was being held—was a relatively simple example of working with the table relationships. Let's look at a more complex example. Suppose you want to find out which room a specific attendee will occupy during a specific timeslot. While the question itself sounds easy, the SQL query must pull information from five separate tables.

Here is a query that finds the room in which the attendee whose last name is 'Shepard' will be between 1 p.m. and 2:30 p.m.

```
SELECT * FROM room
  WHERE room_id IN
  (SELECT room_id
    FROM class, att_class, attendee,timeslot
        WHERE class.session_id =↵
att_class.session_id
      AND att_class.att_id =  attendee.att_id
      AND timeslot.time_id =  class.time_id
      AND attendee.last_name = 'Shepard'
      AND timeslot.time_text = '1 p.m.  -  2:30 p.m.')
```

USING LABELS

Queries like the ones shown above are useful, but they really don't tell us much. It would certainly be helpful if we added some annotation to our "floor plan." For example, using the labeling feature of dbConnect, we could label each room polyline with its corresponding room number, and maybe the capacity of the room. In Chapter 1 you learned how to create labels.

It would also be nice to show the session title and speaker name in each room for a specific timeslot. Unfortunately, with our current linking scheme, this is not possible. The biggest limitation of the labeling feature is that you only have access to the columns in the table to which you are linking. In other words, if you create a label template based on the **ROOM** table, you can only show values from the **room_id**, **room_number**, **capacity**, and **is_lab** columns. Once again, SQL to the rescue!

A Room with a View

To solve this dilemma, you can use a *view*. Instead of linking the rooms (and associated labels) to the **ROOM** table, create a view that includes columns from the **ROOM** table as well as columns from other tables that can provide more meaningful information.

Linking to a view rather than a table has significant advantages, including those listed below:

- It gives you access to columns in related tables for labeling purposes.
- It makes querying multiple tables much easier.
- The view can be based on other dynamic data in your database, allowing you to create more dynamic labels.
- The view itself is easily redefined to include additional tables or columns as your database grows.

The only real disadvantage of linking to a view is performance. Since a view is based on a query, each time you request a linked row or update a label, the database executes the query to create the view, and then AutoCAD queries the view to find the desired row or rows. Linking directly to a table—especially when the linked column is the primary key of the table—is always going to be faster than linking to a view. How much faster depends on the size of the database and the nature of the query that defines the view. A simple query on a relatively small database may result in only a small degradation in performance. If the query is more complex or the database is relatively large, the performance hit may be unacceptable. The only way to determine if linking to a query works for you is to test it using your database.

Let's look at one final example that uses several of the concepts you have learned up to this point. Suppose you want to choose a specific timeslot and have AutoCAD display the room number, session title, and speaker name in each room for that timeslot. Additionally, in the title of the drawing you want to show the textual description of the selected timeslot.

The steps to perform this example are as follows:

1. Create a table called **CUR_TIME** that has one column called **time_id**. This table has only one row in which you will store the **time_id** that you want to represent on the drawing.

2. Create a view called **ROOM_VIEW** that produces the information you want to have in the labels.
3. Create a link template that is based on the view.
4. Create a label template for the labels that are placed in each room.
5. Create a label template for the textual description of the selected timeslot.
6. Create the labels.

Each step is described in detail below. The procedures to accomplish steps 3 through 6 are outlined in a series of short tutorials.

Step 1: Create a Table Called CUR_TIME

The easiest way to do this is to use the facilities included in the database application (in this case, Microsoft Access). In the sample class.mdb file, this table has already been created for you.

Step 2: Create a View Called ROOM_VIEW

Again, the easiest way to do this is to use the query capability of Microsoft Access (in Access, a view is called a *query*). In the sample class.mdb file, this view has already been created for you. It should appear under the class database node in the dbConnect Manager.

Here is the query that is used to create **ROOM_VIEW**:

```
SELECT room.room_id AS datalink, room_number,
session_title, first_name, last_name, time_text
  FROM room, class, speaker, timeslot
  WHERE class.room_id = room.room_id
    AND class.speaker_id = speaker.speaker_id
    AND class.time_id = timeslot.time_id
    AND timeslot.time_id = (SELECT↵
time_id FROM cur_time)
```

If you examine this query carefully, you will find a couple of important things. First, the **room_id** column has been renamed to **datalink**. This was done to distinguish it from the **room_id** column that exists in the **CLASS** table. Second, the last expression in the **WHERE** clause uses a subquery to get the value of the **time_id** column in the **CUR_TIME** table. This is important because it allows the query to produce a completely different result based on that value.

Step 3: Create a Link Template Based on the View

Next, you need to create a link template that is based on **ROOM_VIEW**. The following brief tutorial guides you through this procedure.

TUTORIAL 4.3 – CREATING THE ROOM_VIEW LINK TEMPLATE

1. Open the file room.dwg that you created in the previous tutorial.
2. Open the dbConnect Manager, and connect to the class data source.
3. Right-click the **ROOM_VIEW** view and choose **New Link Template**.
4. Accept the default link template name of **ROOM_VIEWLink1** and click **Continue**.
5. Check the **datalink** column as the key column and click **OK**.

Step 4: Create a Label Template for the Labels Placed in Each Room

Next, you need to create a label template based on **ROOM_VIEW** that will be used to display the desired information within each room. The following tutorial will guide you through this procedure.

TUTORIAL 4.4 – CREATING THE FIRST ROOM_VIEW LABEL TEMPLATE

1. Right-click the **ROOM_VIEW** table and choose **New Label Template**.
2. Accept the default label template name of **ROOM_VIEWLabel1** and click **Continue**.
3. Select the **Properties** tab and change the justification to Middle Center.
4. Select the **Label Fields** tab, choose **room_number** from the **Field** drop-down list and click **Add**.
5. Add the columns **session_title**, **first_name**, and **last_name** using the same process.
6. Click **OK**.

Step 5: Create a Label Template for the Textual Description of the Selected Timeslot

Next, you need to create another label template based on **ROOM_VIEW** that is used to display the textual description of the selected timeslot. The following tutorial guides you through this procedure.

TUTORIAL 4.5 – CREATING THE SECOND ROOM_VIEW LABEL TEMPLATE

1. Right-click the **ROOM_VIEW** table and choose **New Label Template**.
2. Accept the default label template name of **ROOM_VIEWLabel2** and click **Continue**.
3. Select the **Label Fields** tab, select **time_text** from the **Field** drop-down list and click **Add**.
4. Click **OK**.

Step 6: Create the Labels

Finally, labels must be placed in the rooms, and a single label must be created for the time text.

TUTORIAL 4.6 – CREATING THE LABELS

1. In the dbConnect Manager, right-click **ROOM_VIEW** and choose **View Table**.

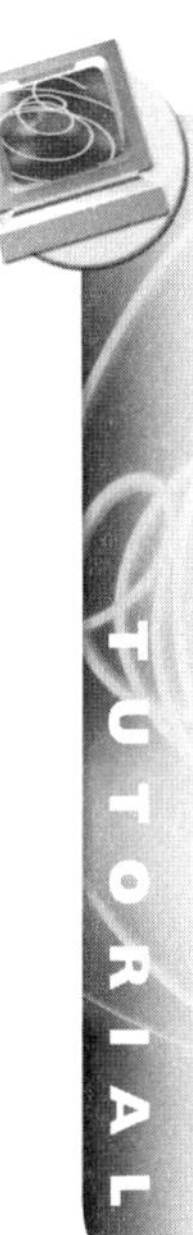

Note: You will probably see only three rows displayed in the Data View window. This is because these are only three rooms that have classes assigned to them during the current timeslot. For the purposes of this tutorial, labeling only the first three rooms will be enough.

2. Make sure that **ROOM_VIEWLink1** and **ROOM_VIEWLabel1** appear in the drop-down lists in the top of the Data View window, as shown in Figure 4.4.

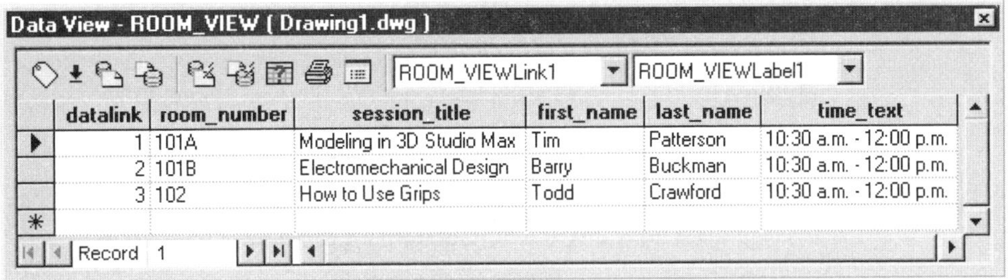

	datalink	room_number	session_title	first_name	last_name	time_text
▶	1	101A	Modeling in 3D Studio Max	Tim	Patterson	10:30 a.m. - 12:00 p.m.
	2	101B	Electromechanical Design	Barry	Buckman	10:30 a.m. - 12:00 p.m.
	3	102	How to Use Grips	Todd	Crawford	10:30 a.m. - 12:00 p.m.
*						

Figure 4.4 *The Data View Window*

3. Click the down arrow next to the **Link** toolbar button and choose the **Create Freestanding Labels** icon.
4. Select the first row in the table.
5. Click the **Label** toolbar button and select a point in the center of the first room.
6. Continue until you have placed labels in the first three rooms.
7. In the **Label Template** drop-down list, choose **ROOM_VIEWLabel2**.
8. Select the first row in the table again.
9. Click the **Label** toolbar button and select a point off to the side, or below the room polylines.
10. Save the drawing.

Creating the Interface

Now that you have established your conference room drawing and have created all the necessary links and labels, the next step is to give users of this drawing an easy way to select a timeslot and update the labels. The following tutorial demonstrates how this is accomplished.

TUTORIAL 4.7 – CREATING THE INTERFACE

1. In the dbConnect Manager, right-click **ROOM_VIEW** (or any table for that matter) and choose **New Query**.
2. Type "Set_Time_1" in the **New query name** box.
3. In the Query Editor, select the **SQL Query** tab.
4. Type the following SQL statement in the text area:
   ```
   UPDATE cur_time SET time_id=1
   ```
5. Click **Store**, and then click **Close** (do not execute it yet).
6. In the dbConnect Manager, right-click the Set_Time_1 query and choose **Duplicate**. This creates a duplicate query called Set_Time_2.
7. Repeat step 6 to create another duplicate called Set_Time_3.
8. Right-click Set_Time_2 and choose **Edit**.
9. Change the **1** to a **2** in the SQL Statement as follows:
   ```
   UPDATE cur_time SET time_id=2
   ```
10. Click **Store** and then click **Close**.
11. Right-click Set_Time_3 and choose **Edit**.
12. Change the **1** to a **3** in the SQL Statement as follows:
    ```
    UPDATE cur_time SET time_id=3
    ```
13. Click **Store** and then click **Close**.
14. Save the drawing.

Making it Work

Displaying the correct sessions and speaker names for a particular timeslot is now just a very simple two-step process:

1. Double-click the query that corresponds to the timeslot you want to view.
2. Right-click the drawing name in the dbConnect Manager, and choose **Reload Labels**

SUMMARY

Since the DbConnect tools use SQL as the primary query mechanism, fluency in SQL is critical to your gaining the most power out of a database that is linked to AutoCAD. This chapter is by no means a complete SQL tutorial. Its purpose is to focus on the features of SQL that are important to AutoCAD users.

You have learned how to

- Compose simple queries using **SELECT**
- Retrieve specific rows using a **WHERE** clause
- Build complex queries using multiple tables
- Understand the concept of the SQL **VIEW**
- Modify a database using **INSERT**, **UPDATE**, and **DELETE**
- Use SQL to check database integrity
- Understand the advantages and disadvantages of linking to a **VIEW**

REVIEW QUESTIONS

1. What are the four levels in the SQL hierarchy?
2. What is an aggregate function?
3. What can the **AS** keyword be used for?
4. What happens when you omit the **WHERE** clause when issuing a **DELETE** command?
5. What is the difference between **WHERE** and **HAVING**? In what order are they evaluated?

EXERCISES

1. Compose SQL statements that check the class database for the following conditions:
   - A speaker is assigned to more than one class during a single timeslot.
   - The number of attendees enrolled in a class exceeds the capacity of the room
2. Compose a query that lists which attendees have enrolled in the most advanced classes.
3. Compose a statement that could be used to remove classes from the database if the enrollment is fewer than 10 people.
4. Compose a query that lists attendees whose classes all take place in one room.

Designing AutoCAD/Database Applications

OBJECTIVES

After completing this chapter, you will be able to

- Understand the importance of custom applications that work with databases
- Choose the best storage mechanism for your non-graphic data
- Identify the type of application you have
- Understand the advantages and disadvantages of various linking schemes
- Work through the process of designing a real-world application

INTRODUCTION

Now that you are familiar with AutoCAD's database connectivity features, you've probably found that there is only so much that can be accomplished with an out-of-the-box application. You have most likely thought of many things that you wish you could do, but the user interface either doesn't allow it or makes it very difficult. Perhaps there are repetitive tasks that you could automate, such as linking or populating a database from the drawing.

The dbConnect interface was designed to be very generic, providing the basic data viewing and linking capabilities. In the context of a real-world application, it quickly falls short. To get AutoCAD and dbConnect to do the specific things you need, some level of customization is required. The next three chapters focus on the knowledge and tools you need to start developing custom applications that use external databases.

Chapter 3 walked you through the process of designing a database, without the notion of connecting it to a graphic interface. New challenges arise when you need to make the database useful from within AutoCAD. In most cases, a well-designed database will fit easily into an application that uses AutoCAD as its graphical interface. However, when you know that AutoCAD will be used as an interface to the database, there are a number of issues that need to be considered *during* the database design process. It is important to understand that database design should be your number one priority when you plan your application. However, some decisions about how your data are organized will be influenced by how the information is represented graphically.

This chapter takes an in-depth look at the issues related to the connectivity of graphic and non-graphic data. As in many technologies, there is always more than one way to solve a particular problem. Choosing the approach that best suits the goals of the application, while minimizing the effort to maintain the integrity of the data, can be critical to the success of a CAD/database application.

Following this chapter are the final three chapters of the book, which focus on the specific *application programming interfaces* (APIs) that are used to develop AutoCAD/database applications.

EXAMPLE ASSET MANAGEMENT APPLICATION

As you work through the topics in this and the next three chapters, you will be taken step by step through the design and development of a fairly robust asset management application. This example application tracks space and people within an office environment. It could easily be expanded to track other assets as well. All of the files for the completed application are on the CD-ROM.

WHY DO YOU NEED A CUSTOM APPLICATION?

There are two primary reasons that additional application development is required to really get the most out of AutoCAD's database connectivity:

- To provide a graphical user interface
- To deal with dual environments

PROVIDING A GRAPHICAL USER INTERFACE

A tool such as dbConnect, while powerful and relatively easy to learn and use, is designed to be everything to everyone. In the context of a real-world application, AutoCAD and dbConnect can only go so far to satisfy the needs of the average user. Quite often, you need to provide a specialized interface that is tailored to users who are not experts in AutoCAD or databases. Such an interface insulates those users from the nuts and bolts of AutoCAD and dbConnect and gives them just the functionality they need to do their specific jobs.

Some examples of custom user interface components include

- Data entry dialog boxes that contain more intelligence and built-in data validation

- Custom query builders that are tailored to specific types of queries
- Custom reports

DEALING WITH DUAL ENVIRONMENTS

Integrating a graphic system and a database system introduces several challenges, which the systems themselves cannot deal with effectively. It is therefore necessary to build additional logic into an application to deal with these challenges. For example, if a user needs to remove an object from the drawing that is linked to the database, a custom application can help ensure that the corresponding row in the linked table is also deleted. Later in the chapter, we will discuss these kinds of problems in more detail.

USING OFF-THE-SHELF APPLICATIONS

Depending on your situation, there may be a suitable off-the-shelf product that uses AutoCAD and databases to perform a specific task. There are numerous commercial products that have been designed for such industries as

- Facilities management
- Parcel mapping
- Municipal and private utilities
- Natural resource management
- Process/Plant design
- Estimating and Scheduling

It would certainly be worth your time to evaluate some existing products to see how well they address your specific problem. If the technology already exists, why reinvent the wheel? Beware, however; many of these products are designed to be all things to all people. Your needs may be simple enough that a commercial product would have more features than you would ever use. Or you might have some specific need that no commercial product can satisfy.

Some of these products may also provide additional customization capabilities, above and beyond what AutoCAD provides. This is certainly an important feature to look for when you evaluate any software product. In any case, whether you use AutoCAD alone or with a third-party product, if you want an application that addresses your specific needs, you are faced with developing some custom applications.

INTERNAL VERSUS EXTERNAL STORAGE OF DATA

You may find it a bit strange that a book about database connectivity would suggest that you don't need to use an external database, but this is a fundamental question you must ask yourself. Connecting external databases to any graphic system means that you must manage and maintain data stored in two completely separate environments. This imposes an additional management burden that you will always be carrying.

You may already have a populated external database that is maintained by some other application but that has information that you want to associate with objects in an AutoCAD drawing. In this case, you don't have much choice. But unless an external database is absolutely necessary or its benefits outweigh the additional effort, it may make more sense not to use it.

On the other hand, storing non-graphic data *inside* an AutoCAD drawing file is not without its difficulties, nor without its limitations. Both data storage methods have advantages and disadvantages. A discussion follows of the issues you will face with each method.

INTERNAL STORAGE

AutoCAD 2000 provides several ways to store non-graphic data inside a drawing file. These methods give users the ability to associate additional information to AutoCAD objects by attaching data directly to the objects. In general, the storage of non-graphic data inside the drawing file can have significant advantages. First, all of the graphic and non-graphic data are contained in a single file, which makes the data easier to manage. Second, since everything is stored in a single file, the objects and the non-graphic data associated with them are virtually inseparable.

Of course, there are disadvantages to this approach. The internal storage mechanisms provide no means to normalize the data. As we learned in Chapter 3, databases generally consist of more than one table in order to store data efficiently and eliminate redundant data. Using the methods described below to store non-graphic data inside the drawing makes this important aspect of database design very difficult, if not impossible. In addition, depending on the volume of non-graphic data you have, sheer file size and performance may be an issue when you store the data in the drawing.

Another disadvantage to internal storage of data is that there are very few tools that allow you to formulate queries on the data. To allow even the simplest query to be performed on the data, you must export the data to a database system.

The two most feasible ways to store non-graphic information inside a drawing file are *block attributes* and *extended entity data*. A third method, which is only available in AutoCAD Map™, is *object data*. These methods are described in detail below.

Block Attributes

Block attributes provide the simplest and most powerful method of attaching non-graphic data directly to AutoCAD objects. Before AutoCAD gave us the ability to link objects directly to external databases, block attributes offered a very feasible way to accomplish a similar task. As database technology evolved, attributes have been considered by many as "old technology," and they are often overlooked. However, they still may be the best solution in many situations. The primary limitations of block attributes: 1) you can only attach attributes to blocks, and 2) the attributes store all data as strings.

The advantage to using attributes is that your data are stored in a very structured fashion. A block that contains attribute definitions can be thought of like a database table structure. Each attribute definition must have a tag name, which could be treated like a column name, and each insertion of a particular block could be treated like a row of values. If an external copy of the attribute data is needed for any reason, you can use the **ATTEXT** command to export the information from the drawing to a delimited text file. The exported file has one line for each block inserted, and a user-defined extraction template file controls which columns (attribute values) are included. The data contained in the delimited text file created by **ATTEXT** can easily be imported to a spreadsheet program or database application. Also, if you know that certain attributes will always contain numeric data (even though they are stored in a character format in the attribute itself) the extraction template file can convert the data to numeric format and even specify the size and number of decimal places.

Another important advantage to using block attributes is that they act just like text objects. They can be created as visible or invisible and manipulated like any other AutoCAD object. When visible, attributes provide a sort of "automatic" annotation feature, similar to labels in dbConnect.

Extended Entity Data

Extended entity data (EED) is another method of attaching non-graphic data to AutoCAD objects. Unlike block attributes, information stored in EED is always invisible. Its biggest advantage over block attributes is that EED can be attached to *any* AutoCAD object, including non-graphic objects such as layers. Another advantage to EED is that it supports a wide variety of data types. The basic data types supported in EED include

- String
- Integer (16 bit and 32 bit)
- Real
- 3D point

In addition, EED supports a number of "smart" data types, including

- Layer names that are automatically updated when a layer is renamed
- Entity handles that are automatically deleted when the corresponding entity is deleted
- Scale factors that are updated when the object is scaled

The biggest disadvantage to EED is that it is totally unstructured. AutoCAD has no built-in controls that ensure that the EED on one object is structured in the same way as that on another object. While EED can be very useful in some situations, it can prove to be very difficult as a substitute for external databases.

On the other hand, applications can be written to allow users to manage EED in a more structured way. A great example of this is the AutoCAD Data Extension (ADE)™. ADE was first available as an add-on to AutoCAD Release 12. More recently, the ADE technology has evolved into the Autodesk GIS product known as AutoCAD Map, and its structured EED functionality has been replaced by a new mechanism called *object data.* Object data, described below, offers yet another internal data storage option, but it's only available in AutoCAD Map.

Object Data (AutoCAD Map Only)

ADE is no longer available as an add-on to AutoCAD. However, all of its functionality is available in AutoCAD Map. AutoCAD Map provides an additional internal data storage mechanism called *object data.* Object data provides the same basic data types as extended entity data. They are

- String
- Integer
- Real
- 3D point

Unlike EED, however, object data is structured more in the way a database is structured. Before you attach object data to an object, you must first create an *object data table.* An object data table is similar to a table definition in an external database. Any number of object data tables can be defined in a single AutoCAD drawing. Each object data table is given a unique name and includes one or more fields. Each field is given a data type, description, and default value.

Once an object data table is created, one or more *records* associated with that table can be attached to any AutoCAD object. Each record consists of a set of values for each field in the table. If the table definition is modified, it affects all the objects that have records attached to them.

AutoCAD Map provides powerful query features that can be used to retrieve objects based on the object data associated with them. It provides routines that allow you to export object data to a delimited text file, or even directly to a database file, while automatically creating the necessary database links on the objects. It also will automatically generate annotation based on object data, or even data in external databases.

EXTERNAL STORAGE

Relational database systems provide a powerful and flexible way to store non-graphic data. You have complete freedom with respect to how your data are organized, and you are able to share your data with other applications much more easily. You are also able to use SQL—the industry standard database language—to formulate queries on your data. Database systems are designed to respond quickly to queries, especially when datasets are very large.

Having a relational database linked to a graphic interface such as AutoCAD gives you the best of both worlds: a powerful graphic editing environment and a flexible, efficient data storage environment. Modifications made in one environment can trigger changes in the other. In addition, results of queries on the database can be displayed within the graphic environment.

Potential Problems

Before you decide that storing your non-graphic data in an external database is the best solution, you must understand some of the issues you will face. There are a number of anomalies associated with integrating a file-based system, such as AutoCAD, with a database system. As you design and implement your AutoCAD/database application, you need to be aware of these potential problems and evaluate how critical they might be in your particular situation.

Data Integrity

When external databases are used to link information to an AutoCAD drawing, you always have at least two files to manage. As long as the graphics and data are in two separate files, there is the potential for them to become out of sync. For example, if you need to maintain an exact one-to-one relationship between the objects and the database, your application must ensure that when an object in the drawing is deleted, the corresponding row in the linked table is also deleted, or at least moved to another table for archiving purposes. Likewise, when additions or deletions are made to the database, those changes need to be reflected in the drawing. There are a number of ways to address data integrity issues within an application, and some examples are demonstrated later in this book.

Transaction Model

In computer lingo, a *transaction* starts when you begin modifying a file or database and ends when the changes are written (permanently) to disk. In a typical session of AutoCAD, you work with an in-memory copy of the drawing file. Changes you make as you work are not written to disk until you explicitly save the file. It is also very easy to make copies of the drawing file to work on. There are times when you might work on a copy of the drawing for hours or even days before you decide whether or not to commit the changes you have made. Also, as a file-based application, AutoCAD has no control over how many copies of a drawing file exist, nor does it know which copy is the "real" one.

This transaction model is vastly different from the way most database systems work. In a database, changes you make are immediately written to the database. Some database systems support longer transactions, but in the database world, *long* could mean just a few seconds.

A transaction-based database allows a rollback of changes at any time prior to those changes being committed. As with AutoCAD, the time between commits can be lengthy, and the user risks losing changes if the system goes down prior to a commit. In such a case, a transaction log can be used to resurrect the lost transactions.

The application must be conscious of these differing transaction models, especially if your application is making changes to the database while the user is working with the linked drawing in a session of AutoCAD. This introduces an obvious problem when the user has made several changes but decides to *quit* the AutoCAD session and not save the drawing file. One way to address this is to force a *save* operation in AutoCAD every time a change is committed to the database—in effect, mimicking the transaction model of the database. In a large drawing, however, this approach may not be acceptable.

An alternative might be to somehow work with a temporary copy of the database table to which you are linking and then transfer all the changes to the "real" table when the drawing is saved. Then again, this could present problems in a multi-user environment. Other users might be working with the same parts of the database at the same time the AutoCAD application is making changes to its "temporary" copy.

Multi-User Support

In a multi-user computing environment, applications must be conscious of which files or databases are in use and appropriately control simultaneous access to those files by multiple users. For file-based applications, such as AutoCAD, this is generally handled by the operating system. For database systems, multi-user access is handled by the database system itself.

It is very important to understand that AutoCAD controls multiple user access very differently than database systems do. AutoCAD uses what is known as *file-level* locking, while databases use *record-level* locking. When a user is working in a drawing file, that entire file is *locked* by the operating system. Other users may be able to *view* a copy of the drawing, but they won't be able to make changes to it. In a database, multiple users can be working with the same database table, and locking only occurs for the brief moment when one user edits a particular row in the table—and only that row is locked.

Security

Security is also handled differently in file-based applications and database systems. File-based applications generally rely on the operating system to handle security, while database systems generally have their own security mechanisms. For example, imagine a user working in a Windows NT® environment and using an AutoCAD drawing that is linked to an Oracle database. As an NT user, he or she might have the appropriate network rights to make changes to the drawing but may not have the same rights in the Oracle system to modify the database.

Dealing with These Problems

All of these problems are the result of working with two independent systems: AutoCAD and a database. If you look at a single system by itself, many of mechanisms to deal with these issues are addressed internally by that system. Take data integrity, for example. If you delete an inserted block that has attributes, AutoCAD takes care of deleting the associated attributes because they are part of the inserted object. In a database system, referential integrity can be established in such a way that deleting a row in one table automatically triggers the deletion of rows in other tables.

When two such systems are combined, the systems themselves have no built-in mechanisms to deal with the problems that are introduced. Ultimately, it is the responsibility of the *application* to deal with these problems and insulate the user from them.

HOW DO YOU DECIDE?

So, do you store your non-graphic data inside the drawing or in an external database? Unfortunately, there is no simple answer to this question. It all comes down to what is important to your application. In order for you to make this decision, there are many factors that you should consider.

Does your data already reside in a database for some other purpose?—While it is ideal to have total control of the database—how it is designed and structured and where it is stored—there are times when you need to connect to a database that already exists and is being used by other applications. In this case, you have no choice but to keep the data external.

Which environment is most critical to your application?—If the integrity of the graphic information is critical to the application, and you can live with a periodic export of the non-graphic data, then storing non-graphic data inside the drawing may be the best solution. If the non-graphic data are more important, and the graphics are simply used as one way of looking at the data, then an external database may be the best approach.

Will you need to manipulate the non-graphic data from applications other than AutoCAD?—You may want to be able to modify the non-graphic data with applications other than AutoCAD. Storing the data inside the drawing would make this impossible, since there are few (if any) database applications that can directly modify data stored in DWG format. You may be able to satisfy the data-sharing requirement using a periodic export of the data from the drawing to an external database. In this scenario, the data would be available to other applications in a read-only form. If the other applications need to modify the data, then the use of external databases may be more appropriate.

How much non-graphic data do you have?—Depending on the amount of data you wish to store, internal storage may not be feasible. Aside from the performance degradation that would result from having a huge drawing file, there are no built-in indexing mechanisms that would speed up a query on internally stored data.

Note: AutoCAD Map provides indexing for object data. The user can select specific tables and fields to index.

What kind of reporting capability will you need?—Tabular reports? Thematic maps? Specialized reporting can be very difficult if the non-graphic data are stored internally. Many database systems provide powerful and flexible reporting tools. For you to take advantage of them, your data must be stored externally.

TYPES OF CAD/DATABASE APPLICATIONS

Okay. You haven't put this book down, so you must have considered the factors described above and have decided that you can best accommodate your application by storing your non-graphic data in an external database. After all, that is the whole idea of this book.

So what kind of application do you have? There are countless applications that can incorporate databases with AutoCAD, and there is no way a single book could cover all the possibilities. There are, however, certain characteristics of AutoCAD/database applications that can be identified, and understanding these can help you choose the right design approach.

Most applications fall into three basic classifications: asset management, bill of materials, and CAD Applications that store application-specific data in a database.

Asset Management

Asset management is probably the most typical AutoCAD/database application, and yet it can be the most complex and challenging. In this type of application, an asset is defined as some sort of tangible object that exists in the real world. Assets are location based, which means they can be represented graphically on a building floor plan or on a map. In asset management applications, the graphical objects (assets) are usually linked directly to the database, and the management of the data can occur either within the graphic environment or within the database. It is therefore imperative that the linkage integrity be maintained and the graphics and database be tightly integrated.

Examples of asset management applications include

- Facilities management (FM) applications, which help keep track of inside- or outside-building assets, such as space, furniture, equipment, electrical and telecommunications networks, and so on
- Maintenance management applications, which track work orders, inspections, and scheduled preventive maintenance on systems or equipment
- Geographic Information Systems (GIS), which are typically used to track outside-building assets, such as utilities, land parcels, bridges, roads and pavement, trees and vegetation, lakes and rivers, air quality, and so on

Having the graphic information linked directly to the database allows the users of these systems to produce graphical reports that were otherwise impossible with just a database. For example, a pavement management application could indicate on a map where road resurfacing is necessary and analyze traffic flow when roads need to be closed. Or an industrial site that uses a map to track the locations of air emissions sources could analyze and display graphically how wind direction and wind speed affect the dissipation of those emissions.

Bill of Materials

A bill of materials is essentially an inventory report of objects that reside, or are represented, in an AutoCAD drawing. The objects in a bill of materials application typically represent real-world objects that are being assembled as part of the design process of a project.

In this type of application, it is less important for the graphical objects to be physically linked to the external database. Instead, the AutoCAD objects may contain additional properties in the form of block attributes or extended entity data. At the user's request, a database is populated from the information in the AutoCAD drawing. A database application can then generate a report, a cost estimate, a purchase order, a bid package, or whatever the specific purpose of the application happens to be.

Another characteristic of bill of materials applications is that they are usually applied earlier in the project cycle. In other words, these applications are used more during the design process as a tool that tracks things as they are being assembled. This also means that the data is extremely dynamic. If the graphic objects are linked to a database, as the designer is working through the various scenarios of a design, data are constantly being added, modified, and removed from the database. This is yet another reason that physically linking the objects during this process may not be desirable, due to the overhead associated with keeping the database in pace with the graphics.

However, once the design phase is completed, and construction (or production) is complete, the associated information could easily be migrated to an asset management system. At that time, the appropriate database links could be established, and the data will be much less dynamic.

Metadata Applications

Metadata applications are applications that store application-specific data in a database. In this type of application, there is no direct relationship between the graphic objects and the information in the database. The database merely contains information *about* the graphic content of the drawing. For example, a company might choose to keep all the information related to its CAD standards in a database and create an interface that helps an AutoCAD user set up a new drawing with standard layers, colors, linetypes, etc.

LINKING SCHEMES

AutoCAD imposes no restrictions on the number of rows that can be linked to an object nor does it restrict the number entities that can be linked to a single row. As a result, you have the flexibility to choose a linking scheme that fits best with the specific needs of your application. There are four possible ways to establish links between AutoCAD objects and external database tables:

- Many-to-one (several objects are linked to a single row in the table)
- One-to-many (each object has links to multiple rows in the table)
- Many-to-many (multiple objects are linked to multiple rows)
- One-to-one (each object is linked to a single unique row in the table)

MANY-TO-ONE OR ONE-TO-MANY

Figures 5.1 and 5.2 illustrate the many-to-one and one-to-many linking schemes. The biggest disadvantage to these linking schemes is that some aspects of the database structure are not known to the database. For example, in a many-to-one scheme, it is impossible to determine from the database how many AutoCAD objects are linked to a particular row.

MANY-TO-MANY

This particular scheme is mentioned only because it is possible—not because it is recommended. If a many-to-many relationship is necessary, then it should not be implemented through the link. Instead, use a one-to-one linking scheme and create the many-to-many relationship within your database.

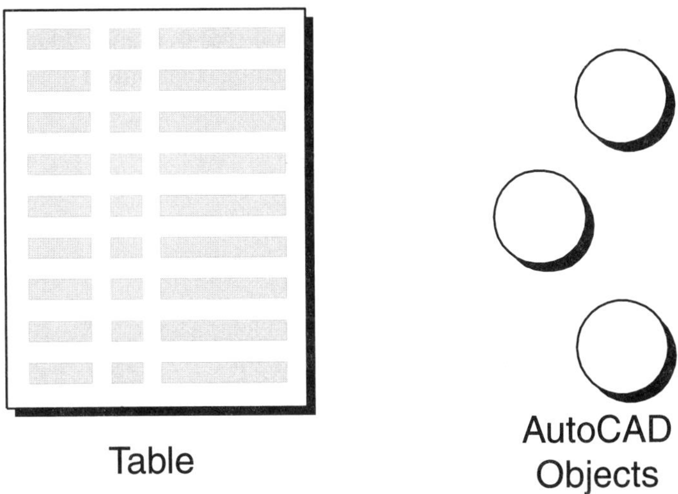

Figure 5.1 *Each Row Linked to Many Objects*

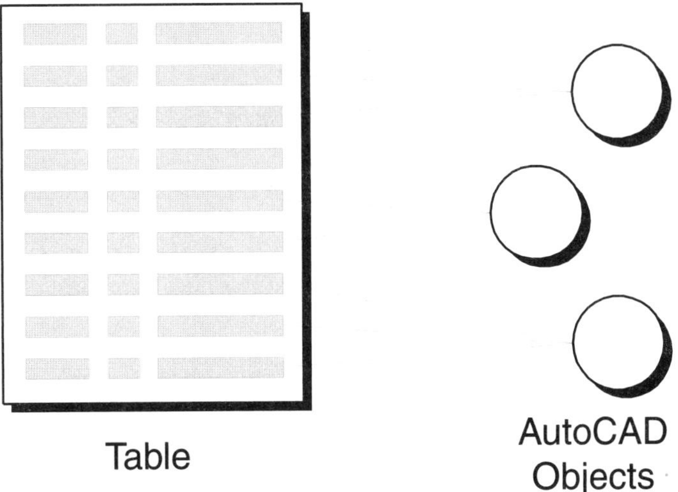

Figure 5.2 *Each Object Linked to Many Rows*

ONE-TO-ONE

The most desirable configuration is a one-to-one relationship (Figure 5.3) between the objects and the table. In this scenario, each object has a single link to a unique row in a table. Ideally, for every row in the table there is a graphical object and for every object there is a row in the table.

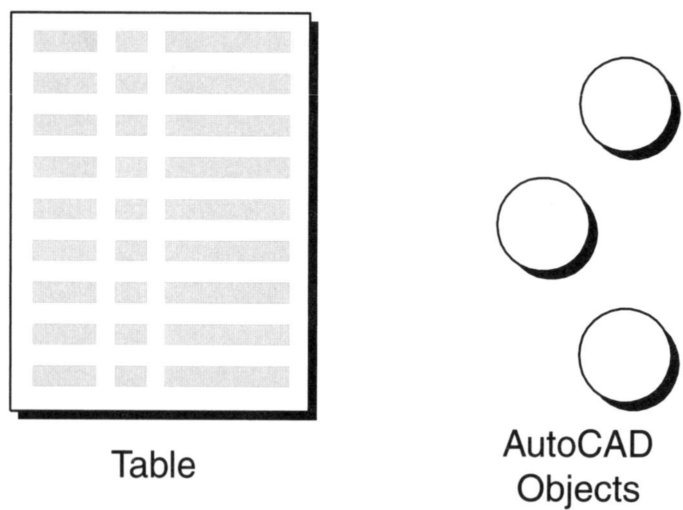

Figure 5.3 *One-to-One Linking Scheme*

In the AutoCAD drawing, you should exercise the same kind of data orderliness as the database imposes on your data. For example, objects that are linked to a common table should all be the same type of object and reside on their own layer. This keeps the graphic information organized and distinguishes the objects that are linked from the ones that are not.

For example, consider a GIS database model in which there is a table of manhole information. A comprehensive spatial database consists of a drawing with every manhole represented as an AutoCAD object, and each manhole object is linked to a unique row in the manhole table. You should be able to gather information about each manhole and modify non-graphic information about a manhole, as well as count the number of manholes that exist in your system, without having to consult the AutoCAD drawing. But as soon as you need to add, delete or move a manhole, you must do this from within AutoCAD.

To maintain the integrity of a true one-to-one scheme, you must reduce the possibility of having orphan rows (rows in your table that do not have an object linked to them) or orphan entities (entities linked to nonexistent rows). Completely eliminating the potentiality of orphans is impossible as long as the graphic and non-graphic data are managed by two independent systems. There is no built-in mechanism that would prevent a user from deleting a row from a database table without also deleting the object to which it is linked. Similarly, there is also no built-in mechanism that would prevent a user from deleting an object from an AutoCAD drawing without also deleting the associated row in the table.

Maintaining the integrity of the links between the graphics and the database then becomes the responsibility of the application. There are many ways your application can reduce the possibility of producing orphans. First, you can provide a specialized interface that is used to manipulate the data from within AutoCAD. Your application would define special commands to add, modify and delete the application-specific objects within AutoCAD and then automatically perform the appropriate database functions. You can also write routines that are run periodically to check the integrity of the database linkages.

GUIDELINES FOR A SUCCESSFUL APPLICATION

The best approach to implementing a successful AutoCAD/database application is to exercise strict organization of data, both within AutoCAD and in the database. This is especially true if you are linking objects directly to rows in database tables. As you design and develop your application, adhering to the following guidelines will help ensure a smooth and successful outcome.

Implement database linkages using a one-to-one scheme—Each object should only contain one link key value that uniquely identifies a single row in the linked table. Your goal should be to minimize, if not eliminate, the need to add, change, or delete link key values on an object. Once the object is linked, the object and its associated data become a single autonomous object. In other words, the linked row of data is an extension of the object, and the object is an extension of the row of data. Maintaining a pure one-to-one linking scheme also means that there should be exactly the same number of linked objects as there are rows in the linked table.

Use only a single column for linking—You should always use the table's primary key for linking purposes. Selecting more than one column for linking can affect performance when you execute queries. Using a single column, or primary key, ensures the best performance possible. If you are using a database that supports indexing, make sure you have established an index for the link column. Indexing allows the system to find a specific row very quickly, without having to iterate through the entire table. If you do not use the primary key, the link column you choose should be restricted in the database to allow only unique values and disallow null values. This will help guarantee that each linked object is associated with a single row.

A link column should exist solely for the purposes of linking—Link columns should be used only to link either to graphic objects or to other tables. The link column should have no other meaningful purpose in the application. If it does, there is always a chance that data in that column might change. Subsequently, the link key value on the object would also need to change. Ideally, the user would never even see the link column and would never have the need, let alone the opportunity, to modify it.

A table should be linked only to a single type of object—As you learned in Chapter 3: Database Design, a table should only contain information that is related to a single type of object. The rules of relational database normalization should apply to the graphic database as well. For example, if you are linking a table of sewer lines to *polyline* objects in AutoCAD, then only link all rows to *polylines*, and do not link *lines* or *arcs*. In addition, it is also good practice to separate the linked objects onto their own layer. This makes it easier to write applications that check for data integrity. It would be very easy, for example, to determine if all the *polylines* on the *sewer line* layer have links on them and that the linked rows exist in the sewer line table.

The table and the object to which it is linked should represent the same type of real-world object—Continuing the theme of the previous guideline, you should also be sure that the object and the table logically represent the same type of object. For example, if you are tracking water service lines and the property addresses to which they are connected, you should not link rows in the address table to the objects that represent service lines. A service line and an address are logically two different types of objects. Instead, you should create a table of service lines and a table of property addresses and establish the relationship between the two tables inside the database. Then link the service line table to the service line object.

You may think this is excessive, especially if you do not plan to track any other information about the service line. Suppose for a moment that you do link the address table to the service line object. If the address changes or the property is split into two addresses, this would require changes to the link key on the object.

DESIGNING AN ASSET MANAGEMENT APPLICATION

To help illustrate some of the concepts presented up to this point, we will walk through the design process of a real-world asset management application. This example application, although simplistic, illustrates many of the common issues you will face as you work through the design and development of your own applications.

The design process used here follows exactly the same process as presented in Chapter 3. Even though the fact that we are connecting to AutoCAD adds some level of complexity to the problem, the design process is still the same. As we develop the entities that make up our database design, the objects in AutoCAD simply become an extension, or a graphic representation, of those entities.

THE PROBLEM STATEMENT

Imagine you have been asked to develop a simple application for the manager of an office facility. To start with, you have the following simple problem statement from a facilities manager:

"I am the facility manager for a medium-sized company, and I need a tool that will help me keep track of the location of employees and departments within my office building."

Your job is to design and develop an application that satisfies the needs of this problem.

The planning approach to this problem will be as follows:

- Determine existing conditions
- Establish the specific requirements of the application
- Design the underlying database
- Identify custom applications

EXISTING CONDITIONS AND NEEDS ASSESSMENT

The first step is to determine existing conditions and gather more detailed requirements from the manager. This is best accomplished through a comprehensive interview and survey process. Interviews should be conducted with all of the primary users of the application. In a large project, it may be necessary to interview several people. In this simplified scenario, we will assume that the only person maintaining and using data with this application is the facility manager. Therefore, we only need a single interview. Described below is an example of how such an interview might occur. The interview is broken into two parts: existing conditions and system requirements.

Existing Conditions

Q. What graphic information currently exists in electronic form?

A. We have a floor plan of the office building in AutoCAD.

Q. What does the floor plan drawing contain?

A. We created the floor plan drawing with the intent of integrating it with a database some time in the future. There are closed polylines that define the perimeter of each space. The polylines are color coded by department. For non-office space, the color is set to "bylayer." Each space is annotated with a text object. For offices and cubicles, the text is the name of the occupant, or "Vacant" if the office is unoccupied. For non-office space, the text describes the use of the space.

Q. What kind of non-graphic data do you have?

A. We do not have any kind of database at this time. We hope to be able to extract some of the information from our floor plan drawing to use as a starting point.

Requirements

Q. What kind of information do you need to track about each employee?

A. I need their name, title, telephone extension, and the department to which they belong.

Q. Is there any specific information you need to track about each department?

A. No, nothing other than the department names—which seem to change every few months for one reason or another.

Q. What kind of information do you need to track about each space?

A. I need the use and/or user of the space, similar to the way the floor plan drawing is currently annotated. I need to be able to quickly see where vacancies exist. I would also like to keep track of which spaces are enclosed, and which spaces have windows.

Q. Are there cases where more than one employee might occupy a single space, such as during second or third shift?

A. Yes, some of the departments have employees working second shift.

Q. In which environment do you anticipate the majority of the maintenance activity occurring, graphic or database?

A. I feel most comfortable in AutoCAD, so that's where I would prefer to do most of my work. For instance, it would be nice to have an object within AutoCAD that represents an employee, such as a text object with that person's name. If that person moves from one location to another, I want to be able to simply move that piece of text to the new location, and have the database automatically updated.

Q. Are you responsible for adding or removing employees from the database?

A. The human resources department maintains the master database of employees. It would be ideal to connect directly to their database, but due to security reasons, I am forced to maintain my own database of employees. The Human Resources department sends me a report each week of additions, modifications, or deletions that have been made.

Q. Do departments "own" particular spaces, or is the departmental assignment of a space simply determined by the department to which the occupant belongs?

A. The departmental assignment of a space is determined by the occupant.

Q. Do you need to track spaces such as storage rooms or conference rooms?

A. Yes. I need to classify every space that could potentially be occupied as an office and indicate the current use of the space.

Q. What about square footage of space?

A. Knowing the area of each space would be nice, but it's not critical.

Q. Do physical changes to the floor plan occur often, such as rearranging partition walls, etc.?

A. No, not very often, maybe once a year.

Q. What kind of reporting capability is needed? Tabular reports? Color-coded floor plan?

A. I need to produce an annotated floor plan that shows each space, who the user of the space is, and what department that user belongs to.

Summary of Interview Process

Once the interview process is complete, it is a good idea to document the primary requirements of the application. For the purposes of our scenario, we will focus on the requirements listed below. These requirements are divided into two categories: data maintenance and data viewing/reporting.

Data Maintenance Requirements

- Track the user and use of every space in the office
- Easily manage employee moves

Data Viewing/Reporting Requirements

- Graphically show space types, department locations, and vacancies
- Annotate the floor plan with employee names and space types

DATABASE DESIGN

The next step is to determine how the data will be organized and how the tables will be linked to the AutoCAD drawing. This is the most important part of the planning process. First, we will establish the primary tables that are absolutely necessary for the application. Then we will start making decisions about which objects will be linked to which tables. Finally, we will fine-tune our database design through normalization.

Establishing Primary Tables

Since the primary focus of the application is to track employee information, a table for employees is necessary. For each employee, we will track the following:

The employee's name—Two columns (one for the first name and one for the last name) will be used to track the name of the employee.

The employee's job title

The telephone extension

The department in which the employee belongs

Figure 5.4 shows a first draft of an **EMPLOYEE** table.

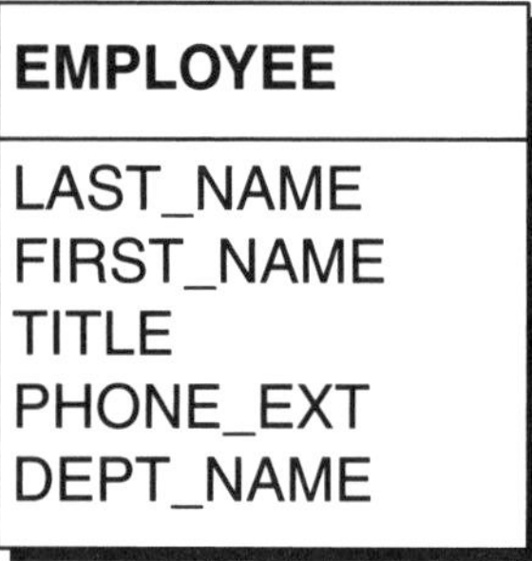

Figure 5.4 *The EMPLOYEE Table*

In addition to the **EMPLOYEE** table, you will need a table that tracks space within the office building. This table will store the following attributes for each space:

The use of the space—This table tracks all types of spaces in the office. Spaces can be cubicle offices, enclosed offices, conference rooms, or even storage closets.

Is the space enclosed?—A yes/no column tells us if the space is enclosed (it has walls and a door) or if it is a cubicle (defined by partition walls).

Does the space have a window?—Another yes/no column indicates if the space has a window.

Figure 5.5 shows our first draft of the **SPACES** table.

SPACES
HAS_WINDOW ENCLOSED USE_TYPE

Figure 5.5 *The SPACES Table*

A relationship between the **EMPLOYEE** table and the **SPACES** table must be established to keep up with which employee occupies which space. This relationship is shown in Figure 5.6. A foreign key called **SPACES_ID** is added to the **EMPLOYEE** table that contains the primary key of the space that each employee occupies. This approach creates a one-to-many relationship between employees and spaces. In other words, an employee can only be assigned to a single space, but a space may have more than one employee assigned to it. This also satisfies the requirement of having more than one employee sharing a single space during different working shifts.

This relationship also allows us to use SQL to determine which offices are vacant and which offices have more than one employee assigned to them.

Figure 5.6 shows the relationship between the **EMPLOYEE** table and the **SPACES** table.

EMPLOYEE
LAST_NAME FIRST_NAME TITLE PHONE_EXT DEPT_NAME SPACE_ID (FK)

SPACES
SPACE_ID HAS_WINDOW ENCLOSED USE_TYPE

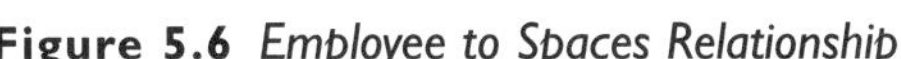

Figure 5.6 *Employee to Spaces Relationship*

Determining How the Database Will Be Linked

Our database design still needs a little bit of work, but before we go much further, we need to start thinking about how the database will be linked to AutoCAD. So far, our database design consists of a table that represents employees and a table that represents spaces. In AutoCAD, each of these object types could easily be represented graphically. We need to determine the best approach to linking—one that both satisfies the needs of the application and minimizes the maintenance effort.

A floor plan of the office will serve as the graphical component of this application. A space could be represented as a closed polygon that defines the perimeter of the space. An employee could be represented as a text object that contains the employee name. Or even better, it could be a freestanding label.

One of the guidelines mentioned earlier states that the table and the object to which it is linked should represent the same type of real-world object. You should not, for example, link rows in the **EMPLOYEE** table directly to the closed polylines in AutoCAD that represent spaces. An employee and a space are two different types of objects. There is already a separate table for spaces that you can use to link to the polylines. And we have already established the relationship between the two tables within the database. This allows us to move employees from cubicle to cubicle without having to change the link on the AutoCAD object.

As you consider your linking options, it is also important to consider what common transactions will be performed on the database and in what environment those transactions will most likely be performed. For example, we know that moving employees from one space to another is one of the most common activities. If users want to do this using direct manipulation of objects inside AutoCAD, then you may need to have an object in AutoCAD that represents an employee. If the user wants to do this exclusively from the database, then an employee object may not be as important.

There will always be more than one way to establish links between a drawing and a database. Even with this simplistic example, there are several approaches that could be taken. Let's examine some of the alternatives.

Link the Spaces Table to Closed Polylines

In this scenario, closed polyline objects are used to represent the individual spaces. In order for a consistent link to be created between the spaces and the database, there needs to be a row for each closed polyline on the floor plan. The advantage to this linking scheme is that, once the links are established, they will hardly ever need to be changed.

Regardless of what other types of objects are linked, the implementation of this scenario is inevitable. It is important to have an object in AutoCAD for each space, because the application needs to track spaces that are not necessarily occupied by an employee.

Link the Employee Table Using Freestanding Labels

In this scenario, each row in the **EMPLOYEE** table is linked to a freestanding label that displays the employee name and could also display other information in that table such as the department name. The advantage to this scheme is that the application can take advantage of AutoCAD's built-in mechanism for updating the labels when the information in the database changes.

There are, however, a couple of disadvantages to this approach. First, it introduces unnecessary duplicate data. If this scheme is implemented, information about which employee occupies which space would exist both within the database and on the drawing. If an employee moves, both the database and the drawing must be modified—a situation that should be avoided if possible. If this approach is implemented, a custom application must be written that synchronizes the drawing and the database.

Another disadvantage to this scheme is that it does not address how unoccupied offices and spaces that are not used as offices are annotated. Some other annotation method would be necessary for these cases, such as standard text objects, or labels that are linked directly to the **SPACES** table. Again, it is preferable to avoid situations where more than one linking scheme is used for a single type of object.

Link Freestanding Labels to a View

In this scenario, all the pertinent information in our database related to a space is consolidated into a view. This view is then linked to freestanding labels placed within each space. The view can give us the occupant of each space and use the **SPACE_ID** as the linking column rather than the **EMPLOYEE_ID**. The advantage to this scheme is that it does not introduce any duplicate data. When an employee is assigned to a new space in the database, the labels are automatically updated.

The problem with this approach is that it assumes that there is only one employee assigned to each space. As we learned earlier, this is not the case. If a view is created that queries information in terms of space, then there is the potential for a space being duplicated in the view. If the **SPACE_ID** were used for linking, then AutoCAD would only find the first occurrence of the space.

Link Only the Spaces, and Programmatically Generate Annotation

In this scenario, only the closed polylines are linked to the SPACES table. Then a custom application is written that creates the appropriate annotation for each space, based on the information in the database. This application essentially bypasses the built-in label functionality that AutoCAD provides. Additional applications could be written that would allow the user to move a text object, representing an employee name, from one space to another and then make the appropriate change to the database. This was one of the wish-list items that was revealed during the interview.

The Conclusion

We know that at least the **SPACES** table will be linked to the closed polylines on the floor plan. The real question is how to handle the annotation. Unfortunately, there is no easy answer. There are two primary factors that contribute to the complexity of this problem:

More than one employee may occupy a single space—This is a very important aspect of our example problem; however, it introduces additional challenges with respect to annotation, such as text placement when two or three employee names are shown in a single space.

The source of the annotation is different depending on the use of the space—If the space is an office, then we want to show the employee name unless the office is vacant, in which case we want to show the word "Vacant." For non-office space, we want to show the type of space.

It seems that no matter how simple a problem may seem, there is always at least one requirement that introduces some sort of unique challenge. This is exactly the reason AutoCAD provides customization capabilities—no single software program can do everything for everyone right out of the box.

While none of the scenarios described above is perfect, with enough custom programming, any of them will work. To make things interesting, we will implement the last scenario, which is to link only the spaces and programmatically manage the annotation (Figure 5.7).

Normalizing the Database

As you learned in Chapter 3: Database Design, in order for data to be stored in the most efficient way, most databases contain several tables. You also learned that the process of fine-tuning the database to eliminate redundancies is called normalization. In our example asset management database, there are a couple of opportunities to further normalize the data.

First, you may want to consider keeping a separate table for departments to avoid duplication in the **EMPLOYEE** table. One of the issues that came up during the interview process was that the department names are occasionally changed. If they reside in the **EMPLOYEE** table, each time the names change, you would have to change every occurrence of the name. If we separate the departmental information into a separate table, changes are made in one place.

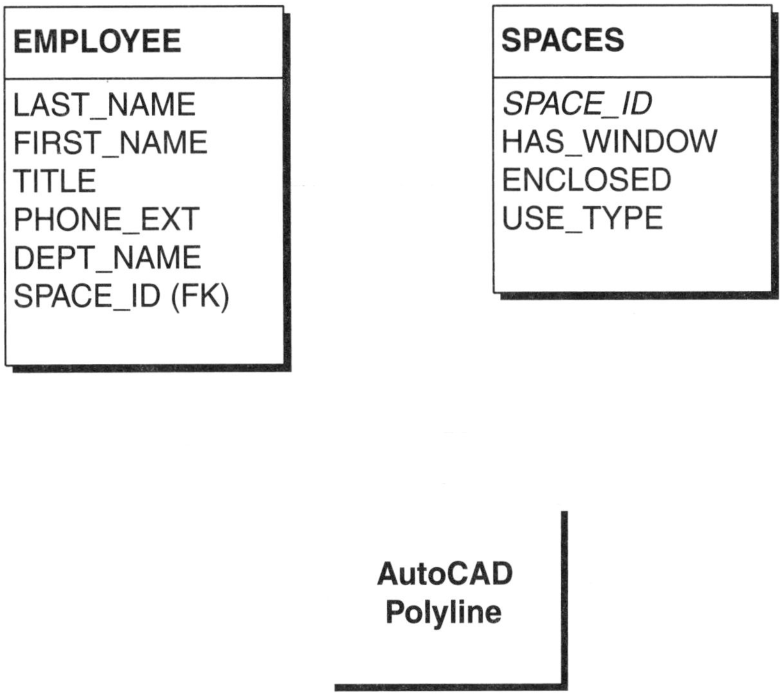

EMPLOYEE
LAST_NAME FIRST_NAME TITLE PHONE_EXT DEPT_NAME SPACE_ID (FK)

SPACES
SPACE_ID HAS_WINDOW ENCLOSED USE_TYPE

AutoCAD Polyline

Figure 5.7 *EMPLOYEE and SPACES Tables with Link to Polyline*

A similar condition exists in the **SPACES** table. The field that designates the type of space has a fixed number of possible values. Subsequently, the space type will be duplicated for each space that has the same type. A table that contains the space types should be added to eliminate the redundant data in the **SPACES** table. Examples of use types:

- Office
- Conference Room
- Storage
- Library
- Plotter
- File Area
- Light Table

After the addition of the **DEPARTMENT** and **USE_TYPES** tables, our database is sufficiently normalized. Figure 5.8 shows the ER diagram of the final database design:

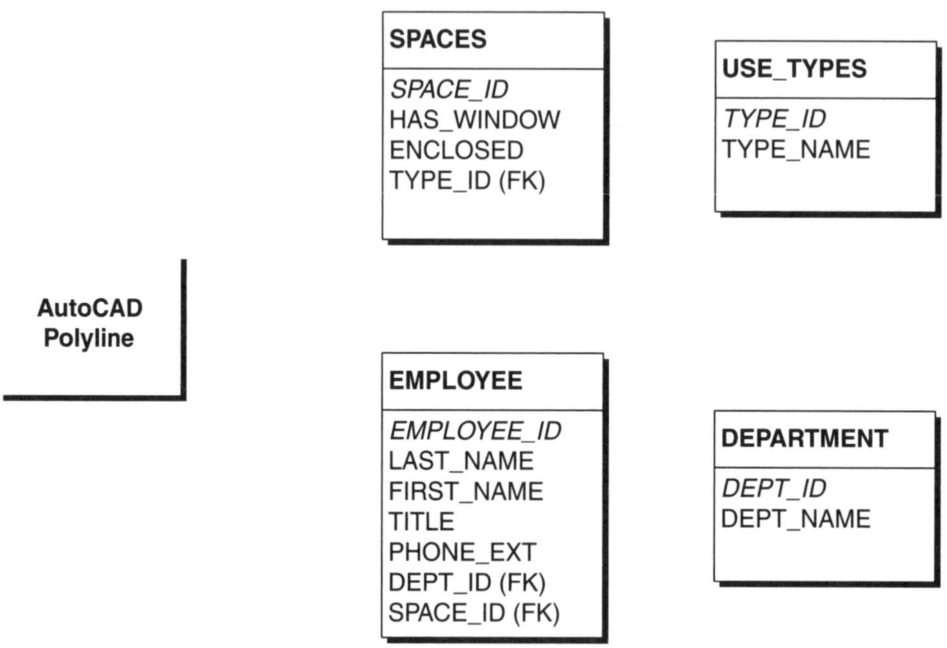

Figure 5.8 *Final Database Design*

IDENTIFYING CUSTOM APPLICATIONS

The last step in our planning process is to identify the specific custom applications we need to develop. At this stage, we only need a brief description of what each application will do. Then as we develop each one, we will produce a more detailed specification.

We will divide the applications into three categories:

- Data Creation
- Data Maintenance
- Reporting

Data Creation

Populate the database from the drawing and generate links—We have a floor plan drawing that contains most of the data we need to begin populating the database. This could be accomplished with a single application that is run once to populate the database. As the application collects the space polylines and writes them to the **SPACES** table, it can also create the link on the polyline object.

Additional applications to populate other columns—in the **SPACES** table, there are two columns that cannot be populated automatically from the drawing. They are **ENCLOSED** and **HAS_WINDOW**. Additional applications will be necessary to help the user populate these columns using graphical selection.

Data Maintenance

Employee additions—An application will be necessary that allows the user to add a new employee to the database and assign that employee to a space.

Employee deletions—Similarly, an application will be necessary that allows the user to delete an employee from the database.

Employee moves—If we use a text object to annotate each space with the employee name, we can create an application that allows the user to physically move the text object from one space to another and have the database update automatically.

Space maintenance—As with employees, applications that provide addition and deletion functions for spaces will be necessary.

Data Integrity Applications—Earlier in this chapter, we discussed some of the issues we face when integrating a CAD environment with a database system. Many of the problems will need to be addressed by our application, such as maintaining the integrity of the one-to-one link between the space polylines and the **SPACES** table.

Reporting

Annotate the spaces—As mentioned earlier, we will develop an application that automatically annotates the drawing according to our own specifications. If the space is an office, then we want to show the employee name unless the office is vacant, in which case we want to show the word "Vacant." For non-office space, we want to show the type of space.

Color-code the floor plan by department—This could be accomplished with hatch objects, because the spaces are represented with closed polylines. A more generic routine could be developed that would allow the user to select the criteria used in the color-coding process.

DEVELOPING THE APPLICATIONS

In order to develop the applications described above, you need to have a firm grasp of one of AutoCAD's development environments such as AutoLISP or Visual Basic for Applications (VBA). In addition, there are specific programming interfaces that must be used to communicate with and link to databases. The remainder of this book covers these interfaces in great detail.

Chapter 6 focuses on Microsoft's ActiveX Data Object (ADO) library. This library is used to communicate with the database. It allows your programs to issue SQL statements and retrieve data from the database. Chapter 7 focuses on Autodesk's programming interface for dbConnect, known as the Database Connection ActiveX Object Model™. Both of these object models are available from both AutoLISP and VBA, and their use in both environments will be demonstrated.

In Chapter 8, we will develop all of the applications necessary to complete our office asset management application.

SUMMARY

Building an application that uses databases with AutoCAD is a complex process. As in any complex process, careful planning and design are necessary to ensure a successful outcome. Be sure you clearly understand what the needs of the application are before you start to create a design. Documenting the requirements of the system will help keep you focused as you start developing any custom applications.

In this chapter, you have learned how to

- Understand the importance of custom applications that work with databases
- Choose the best storage mechanism for your non-graphic data
- Identify the type of application you have
- Understand the advantages and disadvantages of various linking schemes
- Work through the process of designing a real-world application

REVIEW QUESTIONS

1. What methods are available for attaching non-graphic data to an AutoCAD object?
2. What are some of the issues you face when integrating AutoCAD with a database system? Why is it the responsibility of your application to deal with these issues?
3. What is the most desirable linking scheme, and why are the others less desirable?
4. What types of custom applications are typically required in an AutoCAD/database application?

EXERCISES

1. Examine what the office database design would look like if you wanted to track the shifts that people work.
2. Consider adding a table that tracks computer hardware assets in our office application. How would this affect our database design? How would it affect the application?

ActiveX Data Objects (ADO)

OBJECTIVES

After completing this chapter, you will

- Understand the basics of the ADO Object Model
- Be able to use ADO to work with an external database from within an AutoCAD VBA program
- Understand how ADO and other libraries can be accessed from Visual LISP

INTRODUCTION

In Chapter 5 we learned that, in order to accomplish tasks that are tailored to your specific problem, you may need to create a custom application. So where do you begin? AutoCAD has a variety of application programming interfaces (APIs) that can be used, such as Visual Basic for Applications (VBA), Visual LISP and ObjectARX. However, there are additional tools that are necessary for working with databases from within these environments.

It is not the intent of this book to teach the specifics of AutoCAD's APIs. Instead, we will focus on the additional tools that are used within these environments to help you develop custom applications that communicate with, manipulate, and link to external databases. To illustrate how these database connectivity tools are used, we will be working primarily in the VBA environment. Examples of how to use the tools are also demonstrated in Visual LISP. It is beyond the scope of this book to discuss the tools that can be used with ObjectARX.

THE COMPONENT OBJECT MODEL (COM)

It is clear that the trend for programming environments is moving toward the use of Microsoft's Component Object Model (COM) technology. For example, AutoCAD Release 14 exposes its own functionality through a COM interface that allows it to be controlled from other Windows applications. Release 14 also added Visual Basic for Applications (VBA) as a new integrated development environment (IDE). VBA allows programmers to use the AutoCAD object model, as well as object models exposed by other applications.

With AutoCAD 2000, Autodesk continues its API support for COM with an improved object model that includes support for its dbConnect functionality. In addition, AutoCAD 2000 now includes Visual LISP, which gives LISP programmers access to any object model that is accessible from VBA. So regardless of your choice of programming environments, you can take advantage of COM technology.

To develop an application in AutoCAD 2000 that automates database connectivity tasks, you need to be familiar with the following object models:

> **AutoCAD 2000 Objects**—You use these objects in your application to work with entities in the AutoCAD drawing file. While it is beyond the scope of this book to completely cover the AutoCAD object model, there are plenty of other resources available.
>
> **AutoCAD Database Connectivity Automation Objects (CAO)**—You use these objects to work with link templates and the links on the objects in your drawing. This object model is described in detail in Chapter 7.
>
> **Microsoft ActiveX Data Objects (ADO)**—You use ADO to connect to and communicate with your external database. As stated in Chapter 1, the dbConnect features of AutoCAD 2000 are based on Microsoft's OLE DB technology. ADO is the COM interface to OLE DB. This chapter is dedicated to describing ADO in detail.

FILES NEEDED FOR THIS CHAPTER

Rather than get bogged down in the details of implementing the office application just yet, this chapter presents ADO using a more theoretical example. In Chapter 8, we will take what we've learned and apply it to the office example.

The following files are needed to complete the tutorials in this chapter. All of these files are included on the CD-ROM.

> **circles.mdb**—The Microsoft Access database
>
> **circles.dvb**—The AutoCAD VBA project file that contains the sample VBA code
>
> **circles.lsp**—The AutoLISP file that contains the sample AutoLISP code

THE ADO OBJECT MODEL

Figure 6.1 shows a diagram of the basic ADO object model.

Figure 6.2 shows the objects that have a properties collection.

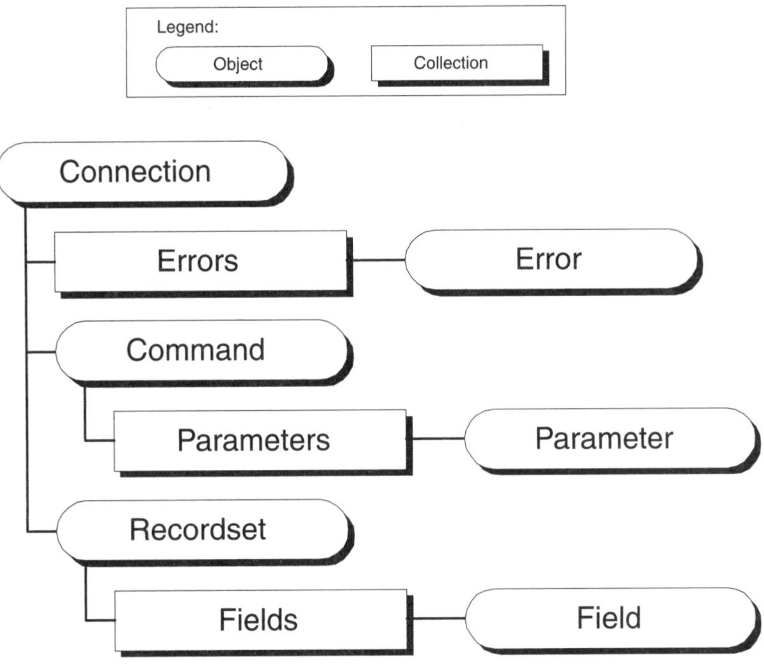

Figure 6.1 *ADO Object Model*

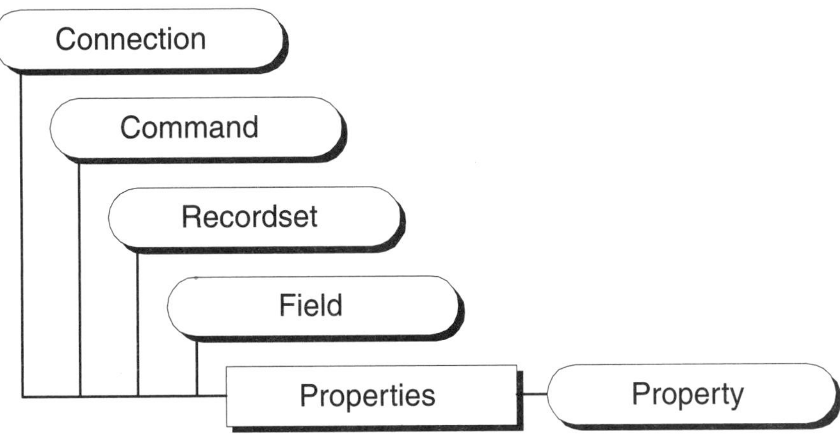

Figure 6.2 *Objects with Properties Collection*

OVERVIEW OF ADO

A TYPICAL DATABASE TRANSACTION

When you need to read from or write to a database from within an application, you will—in very simplistic terms—perform the following steps:

1. Establish a connection to your database.
2. Execute a query or other SQL statements.
3. In the case of a query—where rows are returned—do some manipulation of the data or store it in memory for later use.
4. Close the connection.

Establishing a Connection to Your Database

The ADO **Connection** object handles connections to databases. To establish a connection, you use the **Open** method, which has the following syntax:

connection.**Open** ***ConnectionString*, *UserID*, *Password*, *Options***

The parameters for the **Open** method are as follows:

ConnectionString—A string containing the connection information. This string is typically a list of <parameter>=<value> pairs, separated by semicolons. If you are connecting to a data source that was configured from within AutoCAD, you can use the connection information contained in the UDL file. To do this, use the "File Name" parameter and specify the name of the UDL file as the value like this: "File Name=<*UDL file name*>." (See the example code below.)

UserID—A string containing the user name to use when establishing the connection.

Password—A string containing the password to use when establishing the connection.

Options—The *Options* argument is used to indicate whether you want the **Open** method to wait for the connection to be established or to return immediately (before the connection is established). Many of the methods in ADO support *asynchronous execution*, which means that they return control to your program immediately—before the specified operation has completed. In order for your program to be notified when an asynchronous operation is complete, you must use the corresponding *event*. For example, if you choose to execute the **Open** method asynchronously, you would need to use the **ConnectionComplete** event to determine when the connection is available.

Using the UDL file created by AutoCAD

As you know, in order to link objects to an external database, you must configure a data source in AutoCAD. Doing so creates a UDL file. If you are using ADO to connect to an AutoCAD-configured data source, you can use the information in the UDL file to establish the connection. In VBA, you can get the configured location of the UDL files like this:

```
WsPath =ThisDrawing.Application.Preferences.Files⏎
.WorkspacePath
```

So the connection string needed to connect to the "office" data source is as follows:

```
conString = "File Name=" & WsPath &  "\office.udl"
```

Once you have this, establishing the connection using ADO is easy:

```
Dim db As New ADODB.Connection
db.Open conString
```

Executing a Query

To retrieve data from your database, you use the **Recordset** object. The **Recordset** object has an **Open** method that is used to initialize the **Recordset** object. When you open a **Recordset** object, you can either specify the name of a table or supply an SQL **SELECT** statement. The **Open** method has the following syntax:

> ***recordset*.Open *Source*, *ActiveConnection*, *CursorType*, *LockType*, *Options***

The parameters for the **Open** method are as follows:

> *Source*—Typically a table name or an SQL statement. The *Options* parameter (described below) can be used to tell ADO what this parameter represents.
>
> *ActiveConnection*—A valid **Connection** object variable or a connection string. If this argument is a connection string, then a **Connection** object is not necessary. The **Recordset** object will create its own **Connection** object.
>
> *CursorType*—The type of cursor used when the recordset is opened. This can be one of the following constants:
>
> > **adOpenDynamic**—Opens a dynamic-type cursor. Allows viewing of any additions, deletions, or modifications made to the data by other users while the recordset is open.
> >
> > **adOpenKeyset**—Opens a keyset-type cursor. Similar to a dynamic-type cursor, except that only those modifications to existing data made by other users is visible. Additions and deletions are not available.
> >
> > **adOpenStatic**—Opens a static-type cursor. Creates a static copy of the data. Changes to the data made by other users are not visible.
> >
> > **adOpenForwardOnly**—Opens a forward-only-type cursor. Same as a static-type cursor, except that movement through the cursor is only allowed in the forward direction. This is the default.

LockType—The type of locking used when the recordset is opened. This can be one of the following constants:

adLockReadOnly—The data cannot be modified (default).

adLockPessimistic—Records are locked as soon as you begin modifying data.

adLockOptimistic—Records are locked only when you call the **Update** method.

adLockBatchOptimistic—Only used for batch update mode.

Options – Specifies how the provider should evaluate the *Source* argument. This can be one of the following constants:

adCmdText—The provider should evaluate *Source* as a textual definition of a command.

adCmdTable—ADO should generate an SQL query to return all rows from the table named in *Source.*

adCmdTableDirect—ADO should simply return all rows from the table named in *Source.*

adCmdStoredProc—The provider should evaluate *Source* as a stored procedure.

adCmdUnknown—The type of command in the *Source* argument is not known.

Retrieving the Data

The next step is to step through the cursor and get specific field values. To step through the cursor, you use the **MoveFirst**, **MoveLast**, **MoveNext**, and **MovePrevious** methods provided by the **Recordset** object. You can use the **BOF** and **EOF** properties to determine if you have reached the beginning or end of the cursor respectively.

As you navigate through the cursor, ADO maintains a buffer in memory in which it stores a copy of the *current* record. For field values to be obtained for the current record, the **Recordset** object provides a **Fields** property, which is a collection of **Field** objects. A specific **Field** object can be retrieved by its name or by number with the **Item** property. As a shortcut, a field value can be referenced with an exclamation point, like this:

```
recordset!fieldname
```

The following are examples of obtaining the value of the **LAST_NAME** field from a **Recordset** object called RsEmployee:

```
RsEmployee.Fields(3).Value
RsEmployee.Fields("LAST_NAME").Value
RsEmployee.Fields("LAST_NAME")
RsEmployee!LAST_NAME
```

Notice that the **Value** property of the **Field** object is the default property, so specifying it to retrieve the field value is optional.

Closing the Connection

Both the **Connection** and the **Recordset** objects have a **Close** method that you should call when you are finished using the object. The **Close** method closes the object but does not remove it from memory. To remove the object from memory, set the object variable to **Nothing**.

The following example closes the **Recordset** object called RsEmployee and removes it from memory:

```
RsEmployee.Close
Set RsEmployee = Nothing
```

USING ADO WITH AUTOCAD VBA

REFERENCING THE ADO LIBRARY

In order to use the ADO objects, methods, and properties in an AutoCAD VBA macro, you must *reference* it in the VBA Macro Editor. The tutorial that follows walks you through the steps required to do this.

Note: The following steps must be taken each time you create a new VBA project that uses ADO.

TUTORIAL 6.1 – REFERENCING THE ADO LIBRARY IN VBA

1. Launch AutoCAD 2000 and open a blank drawing.
2. Type "VBAIDE" to launch the VBA Macro Editor.
3. From the **Tools** menu, select **References.** The **References** dialog box appears, as shown in Figure 6.3.
4. Find **Microsoft ActiveX Data Objects 2.0 Library** in the list of available references.

Note: If you do not find the ADO library in your list of references, or the version number is not 2.0 or greater, you can obtain the latest ADO library from Microsoft's Web site. ADO is part of the Microsoft Data Access Components, which can be downloaded from **http://www.microsoft.com/data**.

5. Select the check box beside that item.
6. Click **OK** to close the **References** dialog box.

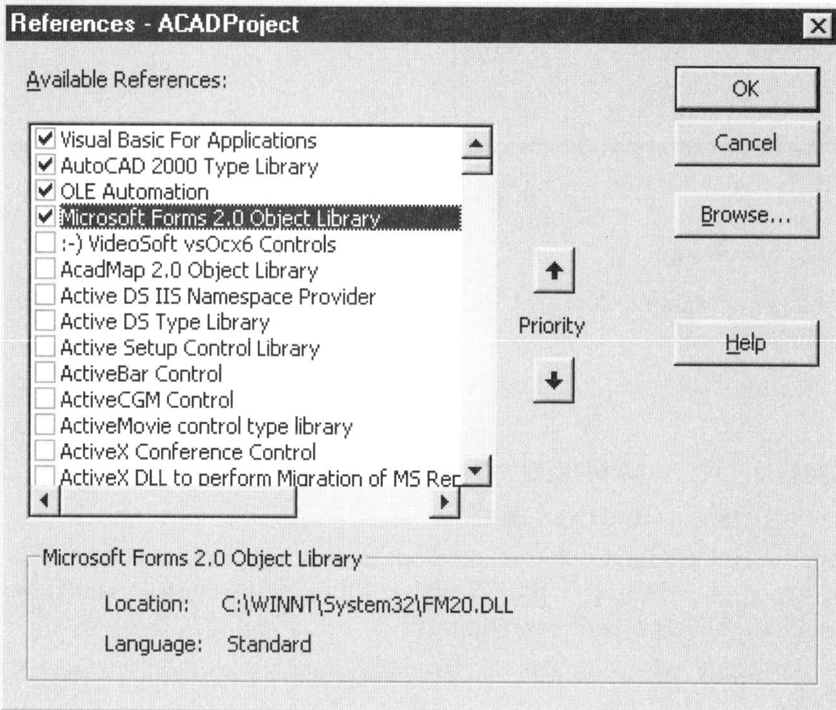

Figure 6.3 *VBA Object References Dialog Box*

ITERATING THROUGH A RECORDSET

When a **Recordset** object is first opened, the record pointer is on the first record. In VBA, the best way to iterate through the recordset is to use the **While/Wend** construct. The **MoveNext** method is used to skip to the next row in the recordset. The **Recordset** object has an **EOF** property (meaning End of File) that can be used in the **While** expression. **EOF** is set to **True** when the end of the recordset is reached.

Shown below is an example AutoCAD VBA macro that lists all the values of the **DEPT_NAME** column from the **DEPARTMENT** table in the office data source.

```
Sub ListDepartments()
  Dim rsDepartment As New ADODB.Recordset
  Dim conString As String

conString = "File Name=" & _
  ThisDrawing.Application.Preferences.Files↩
.WorkspacePath & _
  "\office.udl"
```

```
rsDepartment.Open "DEPARTMENT", conString, _
  adOpenForwardOnly, adLockReadOnly,↵
adCmdTableDirect

While Not rsDepartment.EOF
  ThisDrawing.Utility.Prompt vbCrLf & _
    rsDepartment!DEPT_NAME
  rsDepartment.MoveNext
  Wend

  rsDepartment.Close

End Sub
```

Notice that this particular task is accomplished without the use of a **Connection** object. The second parameter of the **Recordset.Open** method can be either a variable that represents a **Connection** object or the connection string itself. If you provide a connection string, the **Recordset** object creates its own **Connection** object and uses it. In a real application, you may want to create a global **Connection** object that you use for each **Recordset** object you need to open. This prevents your application from having to establish the connection each time you need to retrieve data from the same data source.

MODIFYING DATA IN THE RECORDSET

As you iterate through the recordset, ADO keeps a copy of the current record in a buffer in memory called the *edit buffer*. The current state of the edit buffer can be obtained through the **EditMode** property of the **Recordset** object. Possible values for the **EditMode** property:

adEditNone—No editing operation is in progress.

adEditInProgress—Data in the current record has been modified but not posted to the database.

adEditAdd—The current record was just added but hasn't been posted to the database.

adEditDelete—The current record has been deleted.

In order for data to be modified, the **Recordset** object needs to be updatable. Whether or not a **Recordset** object is updatable depends on how it was created. To determine whether or not a **Recordset** object can be updated, you can use the **Supports** method as shown in the following example:

```
' Make sure the Recordset RS is updatable
If Not RS.Supports(adUpdate) Then
```

```
    MsgBox "Cannot make changes to the recordset."
    Exit Sub
End If
```

To modify data in the current record, simply assign a new value to a field object. As soon as a field has been given a new value, the **EditMode** property is automatically set to **adEditInProgress**. When you are ready to post the changes back to the database, you can call the **Update** method. To cancel the changes, call the **CancelUpdate** method.

ADDING NEW ROWS TO A TABLE

The **Recordset** object has an **AddNew** method that you can use to append a blank record to a table. When the **AddNew** method is called, the new blank record becomes the current record in the edit buffer, and the **EditMode** property is set to **adEditAdd**. Set field values by assigning the values to the **Field** objects. When you are ready to post the changes back to the database, you can call the **Update** method.

The following example adds a new row to the **rsDepartment** recordset:

```
rsDepartment.AddNew
rsDepartment!DEPT_NAME = "Human Resources"
rsDepartment.Update
```

PUTTING ADO TO WORK IN AUTOCAD

EXPORTING DRAWING INFORMATION TO A DATABASE

The following tutorial demonstrates the use of the **AddNew** method in an example VBA macro. This macro shows how you can populate a database using information from a drawing.

TUTORIAL 6.2 – EXPORTING DRAWING INFORMATION TO A DATABASE

In this tutorial, we create a VBA macro that draws a series of circles in a blank drawing and then stores information about each circle in an external database.

Note: If you do not want to take the time to manually transcribe the code for this tutorial, you can find it in the circles.dvb VBA project file included on the CD-ROM. Type "VBALOAD" at the AutoCAD command prompt, and navigate to the location of the circles.dvb file.

First, we need to configure a data source that points to the circles database. The reason for doing this is simply to make it easier to connect to the database using the UDL file.

1. Launch AutoCAD 2000 and open a blank drawing.

2. From the **Tools** menu, select **dbConnect.** This displays the dbConnect Manager.

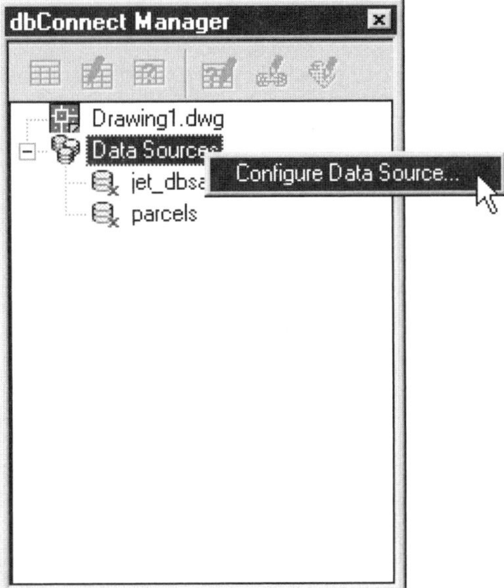

Figure 6.4 *Configure a Data Source*

TUTORIAL

3. Right-click the Data Sources branch in the dbConnect Manager and select **Configure Data Source...** from the shortcut menu.
4. Type "circles" as the data source name, and choose **OK**.
5. Select **Microsoft Jet 3.51 OLE DB Provider** in the list of OLE DB Providers, and choose **Next >>**.
6. Enter the full path of the circles.mdb file.

Tip: You can use the **ellipsis** icon located just to the right of the edit box to bring up a file dialog box.

7. Choose **Test Connection** to verify that a connection can be made.
8. Choose **OK**.

Note: Since we will be connecting to the circles data source from within a VBA program, we do not need to connect to it in the dbConnect Manager.

Next, we create twenty circles in the drawing, with varying sizes and colors. This process is automated with a VBA macro that uses random numbers to generate the circles.

9. Type "VBAIDE" to switch to the VBA integrated development environment (IDE).

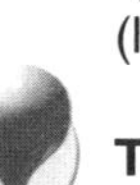

Tip: You can toggle between the VBA IDE and AutoCAD using ALT + F11.

10. Right-click the **VBA Project Explorer** and select **Insert -> Module** from the shortcut menu (Figure 6.5). A code window is displayed.

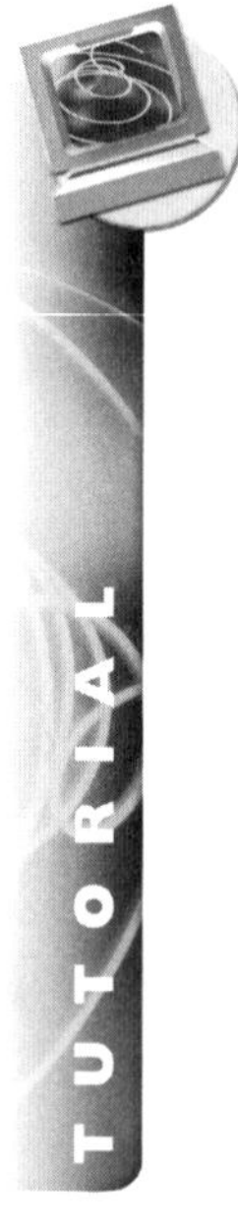

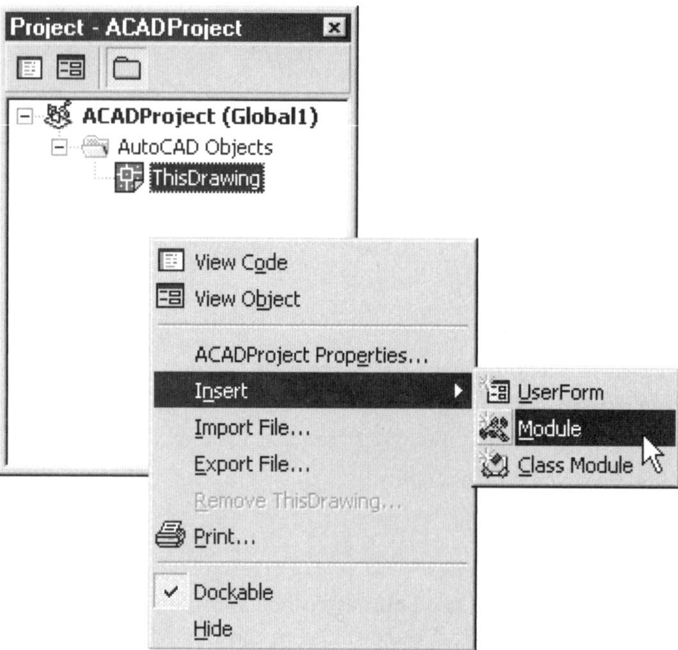

Figure 6.5 *Insert a VBA Module*

11. Type the following code in the code window:

```
Sub CreateCircles()
  Dim newCircle As AutoCAD.AcadCircle
  Dim cen(0 To 2) As Double
  Dim rad As Double
  For i = 1 To 20
    cen(0) = Rnd * 100
    cen(1) = Rnd * 100
    cen(2) = 0#
    rad = Rnd * 20
    Set newCircle = ThisDrawing.ModelSpace.AddCircle↵
(cen, rad)
```

```
    newCircle.Color = Int(Rnd * 7) + 1
  Next
  ThisDrawing.Application.ZoomExtents
End Sub
```

12. Switch back to AutoCAD, type "VBARUN" and press ENTER.

Tip: You can use ALT + F8 to display the **Macros** dialog box.

13. Select **CreateCircles** from the list of macros and click **Run**. You should now see twenty circles in the drawing, with varying sizes and colors.

14. Go back to the VBA Macro Editor, and type the following code in the code window:

```
Sub WriteCirclesToDatabase()
  On Error Resume Next

  ' Variables for AutoCAD Objects
  Dim CircleSelection As  AutoCAD.AcadSelectionSet
  Dim circleObject As AutoCAD.AcadCircle
  Dim groupCode(0) As Integer
  Dim dataValue(0) As Variant

  ' Variables for ADO Objects
  Dim wsPath As String, conString As String
  Dim db As New ADODB.Connection
  Dim circlesRS As New ADODB.Recordset

  ' Connect to the database
  wsPath =↵
ThisDrawing.Application.Preferences.Files↵
.WorkspacePath
  conString = "File Name=" & wsPath &  "\office.udl"
  db.Open conString
  If Err <> 0 Then
    MsgBox "Could not open circles.udl.↵
Make sure " & _
      "it has been configured in the↵
dbConnect Manager."
    Exit Sub
  End If

  ' Open the circles recordset
  circlesRS.Open "CIRCLES", db,↵
adOpenDynamic, adLockOptimistic
```

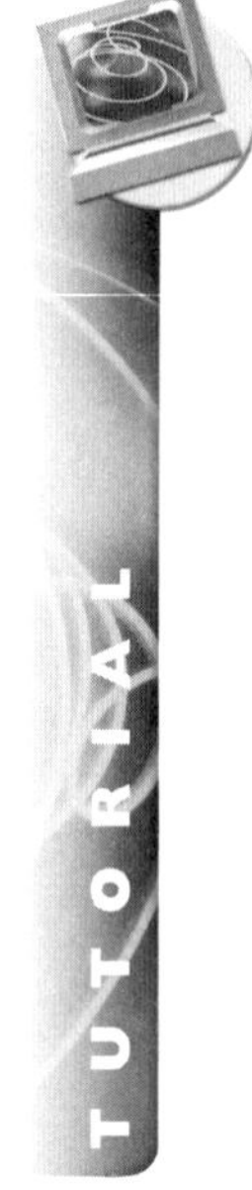


```
  If Err <> 0 Then
    MsgBox "Could not open circles recordset"
    Exit Sub
  End If

  ' Make sure the Recordset supports AddNew
  If Not circlesRS.Supports(adAddNew) Then
    MsgBox "Cannot add records to the  recordset."
    Exit Sub
  End If

  ' Get the selection set of all circles↵
in the current drawing
  Set CircleSelection↵
= ThisDrawing.SelectionSets("Circles")
  If Err <> 0 Then
    Set CircleSelection↵
= ThisDrawing.SelectionSets.Add("Circles")
  End If
  groupCode(0) = 0
  dataValue(0) = "Circle"
  CircleSelection.Clear
  CircleSelection.Select acSelectionSetAll,↵
, , groupCode, dataValue

  ' Loop through the selection and add the↵
circle data to the database
  For Each circleObject In CircleSelection

    ' Add a new blank record
    circlesRS.AddNew

    ' Set the field values
    circlesRS!Handle = circleObject.Handle
    circlesRS!Center_X = circleObject.Center(0)
    circlesRS!Center_Y = circleObject.Center(1)
    circlesRS!Radius = circleObject.Radius
    circlesRS!Color = circleObject.Color

    ' Commit the changes
    circlesRS.Update
        Next
```

```
    ' Close the recordset and the database  connection
    circlesRS.Close
    db.Close
End Sub
```

15. Switch back to AutoCAD, type "VBARUN" and press ENTER.
16. Select **WriteCirclesToDatabase** from the list of macros and click **Run**.

To check the results of the macro, you can view the table in the Data View window.

17. Double-click the circles data source to connect to the database.
18. Double-click the **CIRCLES** table to bring up the Data View window. It should look something like Figure 6.6.

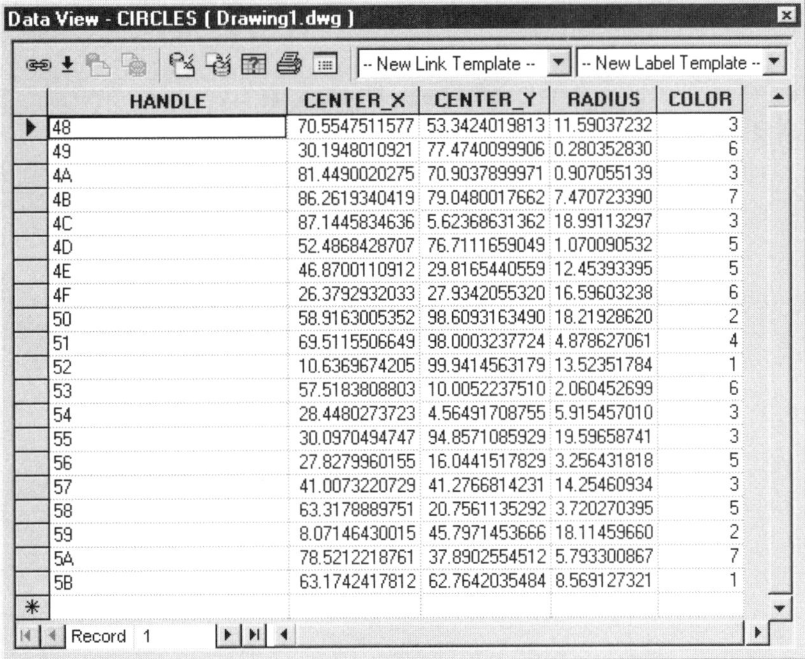

Data View - CIRCLES (Drawing1.dwg)

-- New Link Template -- | -- New Label Template --

HANDLE	CENTER_X	CENTER_Y	RADIUS	COLOR
48	70.5547511577	53.3424019813	11.59037232	3
49	30.1948010921	77.4740099906	0.280352830	6
4A	81.4490020275	70.9037899971	0.907055139	3
4B	86.2619340419	79.0480017662	7.470723390	7
4C	87.1445834636	5.62368631362	18.99113297	3
4D	52.4868428707	76.7111659049	1.070090532	5
4E	46.8700110912	29.8165440559	12.45393395	5
4F	26.3792932033	27.9342055320	16.59603238	6
50	58.9163005352	98.6093163490	18.21928620	2
51	69.5115506649	98.0003237724	4.878627061	4
52	10.6369674205	99.9414563179	13.52351784	1
53	57.5183808803	10.0052237510	2.060452699	6
54	28.4480273723	4.56491708755	5.915457010	3
55	30.0970494747	94.8571085929	19.59658741	3
56	27.8279960155	16.0441517829	3.256431818	5
57	41.0073220729	41.2766814231	14.25460934	3
58	63.3178889751	20.7561135292	3.720270395	5
59	8.07146430015	45.7971453666	18.11459660	2
5A	78.5212218761	37.8902554512	5.793300867	7
5B	63.1742417812	62.7642035484	8.569127321	1

Record 1

Figure 6.6 *The* **CIRCLES** *Table in the Data View Window*

19. Save the drawing as circles.dwg.

TUTORIAL

UPDATING THE DRAWING FROM THE DATABASE

To continue our previous example, suppose we want to update the circles in the drawing according to the current values in the database. You probably noticed that one of the data items we stored was the *handle* of each circle. This represents a somewhat crude but effective way of establishing a link between the object and the row in the table, without using the dbConnect linking mechanism (we'll get to that in the next chapter).

So the process is as follows:

1. Read each row from the database.
2. Find the specific circle using the stored handle.
3. If the circle is found, modify its properties according to the current values in the table.
4. If a matching circle object cannot be found, it is created and its new handle is stored in the database.

TUTORIAL 6.3 – UPDATING THE DRAWING FROM THE DATABASE

1. Open the circles.dwg you created in the previous tutorial.
2. Switch to the VBA IDE.
3. Double-click the module you created to open the code window.
4. Type the following code in the code window:

```
Sub ModifyCirclesFromDatabase()
  On Error Resume Next

  ' Variables for AutoCAD Objects
  Dim circleObject As AutoCAD.AcadCircle
  Dim cen(0 To 2) As Double
  Dim rad As Double

  ' Variables for ADO Objects
  Dim wsPath As String, conString As String
  Dim db As New ADODB.Connection
  Dim circlesRS As New ADODB.Recordset
  ' Connect to the database
  wsPath =ThisDrawing.Application.Preferences.Files↵
.WorkspacePath
  conString = "File Name=" & wsPath & "\circles.udl"
  db.Open conString
  If Err <> 0 Then
```

```
    MsgBox "Could not open circles.udl.↵
Make sure " & _
       "it has been configured in the↵
dbConnect Manager."
    Exit Sub
  End If

  ' Open the circles recordset
  circlesRS.Open "CIRCLES", db,↵
adOpenDynamic, adLockOptimistic
  If Err <> 0 Then
    MsgBox "Could not open circles recordset"
    Exit Sub
  End If

  While Not circlesRS.EOF
    ' Set cen variable to current database values
    cen(0) = circlesRS!Center_X
    cen(1) = circlesRS!Center_Y
    cen(2) = 0#

    Err.Clear
    Set circleObject = ThisDrawing.HandleToObject↵
  (circlesRS!Handle)
    If Err = 0 Then
      ' Circle was found - set↵
properties to current values
      circleObject.Center = cen
      circleObject.Radius = circlesRS!Radius
      circleObject.Color = circlesRS!Color
      circleObject.Update
    Else
      ' Circle was not found. Create↵
new circle using current values
      rad = circlesRS!Radius
      Set circleObject = ↵
ThisDrawing.ModelSpace.AddCircle(cen, rad)
      circleObject.Color  = circlesRS!Color

      ' Store the handle of the new  circle
      circlesRS!Handle  = circleObject!Handle
      circlesRS.Update
    End If
    circlesRS.MoveNext
  Wend
```

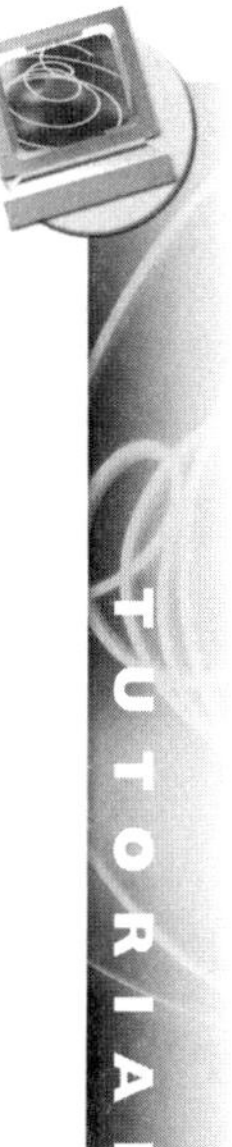

```
    ' Close the recordset and the database connection
    circlesRS.Close
    db.Close
End Sub
```

5. Switch back to AutoCAD.
6. Modify some of the circle objects in some way, such as changing their color, changing their location, or changing their radius.
7. Type "VBARUN" and press ENTER.
8. Select **ModifyCirclesFromDatabase** from the list of macros and click **Run**. The circles are reset to their original state.

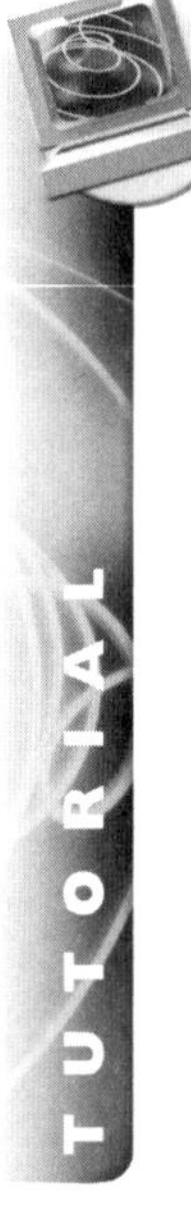

SEARCHING FOR SPECIFIC ROWS IN THE RECORDSET

Using the same example, suppose that instead of updating the drawing according to changes in the database, you want to update the database according to changes in the drawing. In this case you need to perform the following steps:

1. Look at each circle object.
2. Search the database for the corresponding row.
3. Update the database with the current properties of the circle object.

To perform the search operation, you use the **Find** method of the **Recordset** object. The **Find** method has the following syntax:

recordset.**Find** ***Criteria***, ***SkipRows***, ***SearchDirection***, ***Start***

The parameters for the **Find** method are as follows:

Criteria—A string containing a column name, comparison operator, and value to use for the search. This is similar to the text following the **WHERE** clause in an SQL statement.

SkipRows—The offset from the current row or *bookmark* to begin the search.

SearchDirection—Specifies the direction of the search. Possible values are **adSearchForward** or **adSearchBackward**.

Start—A bookmark to use as the starting position for the search. The **Recordset** object has a bookmark property that can be saved for later use at any time. ADO also defines the following bookmarks:

adBookmarkCurrent—Start at the current record.

adBookmarkFirst—Start at the first record.

adBookmarkLast—Start at the last record.

TUTORIAL 6.4 – UPDATING THE DATABASE FROM THE DRAWING

1. Open the circles.dwg you created in the previous tutorial.
2. Switch to the VBA IDE.
3. Double-click the module you created to open the code window.
4. Type the following code in the code window:

```
Sub ModifyDatabaseFromCircles()
  On Error Resume Next

  ' Variables for AutoCAD Objects
  Dim CircleSelection As AutoCAD.AcadSelectionSet
  Dim circleObject As AutoCAD.AcadCircle
  Dim groupCode(0) As Integer
  Dim dataValue(0) As Variant

  ' Variables for ADO Objects
  Dim wsPath As String, conString As String
  Dim db As New ADODB.Connection
  Dim circlesRS As New ADODB.Recordset

  ' Connect to the database
  wsPath =
ThisDrawing.Application.Preferences.Files↵
.WorkspacePath
  conString = "File Name=" & wsPath & "\circles.udl"
  db.Open conString
  If Err <> 0 Then
    MsgBox "Could not open circles.udl.↵
Make sure " & _
      "it has been configured in the dbConnect↵
Manager."
    Exit Sub
  End If

  ' Open the circles recordset
  circlesRS.Open "CIRCLES", db,↵
adOpenDynamic, adLockOptimistic
  If Err <> 0 Then
    MsgBox "Could not open circles recordset"
    Exit Sub
  End If
```

```
  ' Make sure the Recordset is updatable
  If Not circlesRS.Supports(adUpdate) Then
    MsgBox "Cannot update the recordset."
    Exit Sub
  End If

  ' Get the selection set of all circles↵
in the current drawing
  Set CircleSelection↵
= ThisDrawing.SelectionSets("Circles")
  If Err <> 0 Then
    Set CircleSelection↵
= ThisDrawing.SelectionSets.Add("Circles")
  End If
  groupCode(0) = 0
  dataValue(0) = "Circle"
  CircleSelection.Clear
  CircleSelection.Select acSelectionSetAll,↵
, , groupCode, dataValue

  ' Loop through the selection and update or
  ' add the circle data to the database
  For Each circleObject In CircleSelection

    circlesRS.Find "HANDLE='" &↵
circleObject.Handle & "'", , , _
      adBookmarkFirst

    If circlesRS.EOF Then
      ' Not found. Add a new blank record
      circlesRS.AddNew
    End If

    ' Set the field values
    circlesRS!Handle = circleObject.Handle
    circlesRS!Center_X = circleObject.Center(0)
    circlesRS!Center_Y = circleObject.Center(1)
    circlesRS!Radius = circleObject.Radius
    circlesRS!Color = circleObject.Color

    ' Commit the changes
    circlesRS.Update
  Next
```

```
  ' Close the recordset and the database connection
  circlesRS.Close
  db.Close
End Sub
```

GETTING THE LIST OF AVAILABLE DATA SOURCES

Obtaining a list of the available data sources in AutoCAD 2000 (those that appear in the dbConnect Manager) is really quite simple. In Chapter 1 you learned that AutoCAD 2000 uses UDL files to store information about how it should connect to certain databases. These UDL files are stored in a central location, which is configurable in AutoCAD. In VBA, the configured directory location of the data sources can be retrieved with the **Preferences** object as follows:

```
ThisDrawing.Application.Preferences.Files↵
.WorkspacePath
```

The data sources that appear in the dbConnect Manager are nothing more than the list of the UDL files that exist in this directory. Using the Visual Basic Dir function, we can find all the files that match the pattern "*.udl" in a specified directory.

The following VBA function creates a collection object that contains the data source names:

```
Public Function getDsList() As Collection
  Dim dsCol As New Collection
  Dim dbfile As String
  Dim wsPath As String

  wsPath =
ThisDrawing.Application.Preferences.Files↵
.WorkspacePath
  dbfile = Dir(wsPath & "\*.udl", vbNormal)

  While dbfile <> ""
    dsCol.Add (Left$(dbfile, Len(dbfile) - 4))
    dbfile = Dir
  Wend

  Set getDsList = dsCol

End Function
```

This handy function can then be used like other collection objects in a **For Each** loop. Here's how it might be used in a form's **Initialize** event, to populate a **List Box** object:

```
Private Sub UserForm_Initialize()
  For Each dsName In getDsList()
```

```
    ListBox1.AddItem dsName
  Next
End Sub
```

GETTING THE LIST OF AVAILABLE TABLES IN A DATA SOURCE

Another common task is to get a list of the tables that are available for a connected data source. If you remember from the discussion of the SQL hierarchy in Chapter 4, the tables in a database are specified in the database's *schema*. In ADO, getting information from the schema is accomplished with the **OpenSchema** method of the **Connection** object.

The following VBA function creates a collection object that contains the table names for a connection:

```
Public Function getTableList(db As↵
ADODB.Connection) As Collection
  Dim tblCol As New Collection
  Dim tblName As String
  Dim objRec As ADODB.Recordset

  Set objRec = db.OpenSchema(adSchemaTables)
  While Not objRec.EOF
    If objRec!TABLE_TYPE = "TABLE" Then
      tblCol.Add objRec!TABLE_NAME
    End If
    objRec.MoveNext
  Wend

  objRec.Close
  Set getTableList = tblCol
End Function
```

EXECUTING SQL COMMANDS

If you need to execute an SQL statement that does not return a recordset, such as **INSERT**, **UPDATE** or **DELETE**, you can use the **Execute** method of the **Connection** object. The syntax of the **Execute** method is as follows:

connection.**Execute** ***Command*, *RecordsAffected*, *Options***

The parameters for the **Execute** method are as follows:

Command—Can be a table name or an SQL statement. The *Options* parameter can be used to tell ADO what this parameter represents.

RecordsAffected—A long integer set by the **Execute** method that contains the number of rows affected by the SQL statement.

Options—Specifies how the provider should evaluate the *Command* argument. The possible values for this parameter are the same as those described earlier in the **Open** method of the **Recordset** object.

The **Execute** method always returns a **Recordset** object. If the SQL statement is a non–row-returning query, such as **INSERT**, **UPDATE** or **DELETE**, there is no need to assign the return value to a variable. If the SQL statement is a **SELECT** statement, then the **Execute** method returns an open **Recordset** object containing the results of the query. In this case, you should assign this return value to a variable.

In our earlier example macro called WriteCirclesToDatabase(), suppose you want to clear the entire circles table before adding the new circles. The SQL statement to delete all the rows in a table is

```
DELETE FROM CIRCLES
```

You could insert the following line of code just after the connection is established:

```
db.Execute "DELETE FROM CIRCLES"
```

USING ADO WITH VISUAL LISP

If you are an AutoLISP programmer, you have probably explored the Visual LISP environment in AutoCAD 2000. Visual LISP supports COM, which means that you can create objects and access their methods and properties just the way you do in VBA. Since ADO is a COM library, all of the functionality of ADO is accessible from Visual LISP.

This section provides you with general functions that can come in very handy when you access databases from within a Visual LISP program. All of the code in this section can be found in the file circles.lsp on the CD-ROM.

ACCESSING COM LIBRARIES

Before you can use any of the COM functions within Visual LISP, you must first call the (vl-load-com) function. If you have a Visual LISP module that uses COM, you can simply include the following line at the beginning of the source file:

```
(vl-load-com)
```

Visual LISP provides several useful functions for accessing COM libraries. Table 6.1 lists just some of these functions:

Let's look at an example of how some of these functions can be used. The following code obtains the **Preferences** object of the current AutoCAD session:

```
(vlax-get-property (vlax-get-acad-object)↵
"Preferences")
```

We can take this a step further to obtain the path to the data source UDL files:

```
(setq wsPath
  (vlax-get-property
    (vlax-get-property
      (vlax-get-property (vlax-get-acad-object)↵
"Preferences")
```

Visual LISP Function	Description
(vlax-create-object *prog-id*)	Creates an instance of a COM object
(vlax-get-acad-object)	Retrieves AutoCAD's top level application object
(vlax-invoke-method *obj method list*)	Calls the specified method of an object
(vlax-get-property *obj property*)	Retrieves the value of a property
(vlax-put-property *obj property arg*)	Sets the value of a property
(vlax-release-object *obj*)	Releases a COM object
(vlax-import-type-library ...)	Creates new Visual LISP functions based on the methods, properties, and constants contained in a COM library

Table 6.1 *Visual LISP Functions Used with COM Libraries*

```
            "Files"
          )
          "WorkspacePath"
        )
      )
```

Earlier, we showed the VBA code to obtain the same information:

```
ThisDrawing.Application.Preferences.Files↵
.WorkspacePath
```

As you can see, using COM libraries in Visual LISP is a bit more complex than it is in VBA. This is because COM was designed with languages like Visual Basic in mind, not Visual LISP.

IMPORTING THE ADO LIBRARY

The easiest way to use the ADO library in Visual LISP is to use the (vlax-import-type-library) function. The following code accomplishes this:

```
(vlax-import-type-library
  :tlb-filename
"C:\\Program Files\\Common↵
Files\\system\\ado\\msado15.dll"
  :methods-prefix "ado-"
  :properties-prefix "ado-"
```

```
        :constants-prefix "ado-"
      )
```

This creates a set of Visual LISP functions that are based on the methods, properties, and constants in the ADO library. With the code above, all of the new ADO functions start with "ado-." For example, in VBA you opened a connection (described by **conString**) with the following code:

```
db.Open conString
```

In Visual LISP the code looks like this:

```
(ado-Open db conString "" "" ado-adOptionUnspecified)
```

Note: In VBA, you could leave out optional parameters. When you invoke methods in Visual LISP, **all** parameters must be supplied. If the number of parameters is incorrect, an error will occur.

Using the Apropos Window to View Function Names

In the Visual LISP editor, you can view a list of available defined functions that match a particular pattern. This is very useful when used in connection with the (vlax-import-type-library) function. To activate the Apropos window, choose Apropos Window from the View menu.

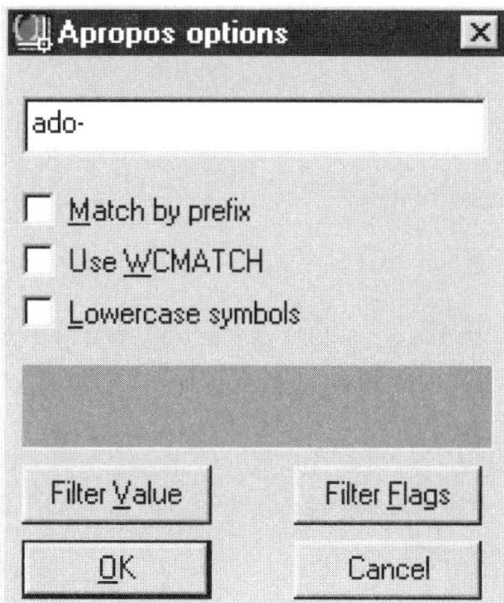

Figure 6.7 *The Apropos Options Window*

To view the ADO functions created by the call to (vlax-import-type-library) above, type "ado-" in the Apropos Options window as shown in Figure 6.7. When you click OK, a window is displayed that lists all available Visual LISP functions that have "ado-" in their name. Figure 6.8 illustrates what this window looks like.

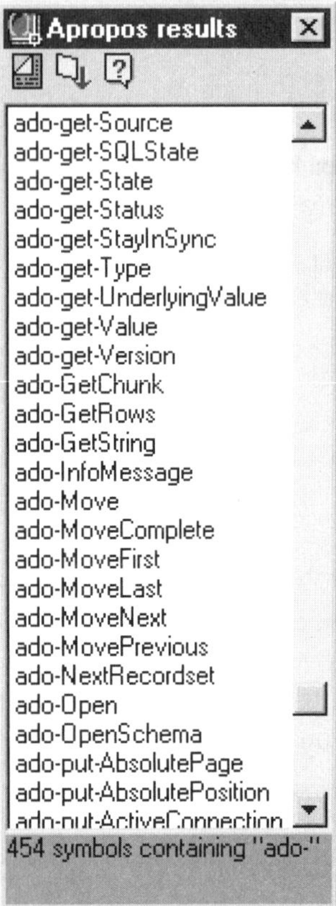

Figure 6.8 *The Apropos Results Window*

Note: The (vlax-import-type-library) function must be executed before the functions become available in the apropos window.

Notice that without the "ado-" prefix, the ADO methods are exactly the same as they are in VBA. For example, the **MoveNext** method is now defined as a Visual LISP function called (ado-MoveNext).

Each of the ADO properties has a **get** function and, unless it is a read-only property, it also has a **put** function. For example, to retrieve the value of the **State** property, you use the function called (ado-get-State).

The ADO constants are also defined. For example, the **adOpenForwardOnly** constant is defined in Visual LISP as ado-adOpenForwardOnly.

When you use method or property functions created by (vlax-import-type-library), the first argument is always the object to which the method or property applies.

Additional ADO arguments required by the method or property start at the second argument.

CREATING AN INSTANCE OF AN ADO OBJECT

To create an instance of an ADO object in Visual LISP, you use the (vlax-create-object) function. If successful, the (vlax-create-object) function returns a pointer to the new object. If it fails, nil is returned.

For example, to create a new **Connection** object and store it in a variable called **db**, you could use the following line of code:

```
(setq db (vlax-create-object "ADODB.Connection"))
```

In a typical application, you might want to maintain a global **Connection** object that is used throughout. Shown below is a simple, yet handy function you could use to set or return a global **Connection** object. If the object creation fails for whatever reason, an alert box notifies you, and the function returns nil.

```
(defun adoDbConnection()
  (cond
    (*adoDbConnection*)
    ((setq *adoDbConnection*↵
(vlax-create-object "ADODB.Connection")))
    ((alert "Could not create↵
ADODB.Connection object."))
  );cond
);defun
```

Then any time the **Connection** object is needed in your application, you simply specify a call to this function as follows:

```
(adoDbConnection)
```

ERROR TRAPPING

Under normal circumstances, if an error occurs when you use a COM library function, Visual LISP calls the defined (*error*) function and exits the program. If you want to have more control over error handling within your program, Visual LISP provides the (vl-catch-all-apply) function. The syntax for this function is as follows:

```
(vl-catch-all-apply function argumentList)
```

This function works much like the (apply) function in AutoLISP. For example, the following code, which appeared earlier in the chapter, is used to open a **Connection** object.

```
(ado-Open db conString "" ""↵
ado-adOptionUnspecified)
```

The same operation using the (vl-catch-all-apply) function looks like this:

```
(vl-catch-all-apply 'ado-Open
  (list db conString "" "" ado-adOptionUnspecified)
  )
```

If the **Open** method returns an error, the (vl-catch-all-apply) function returns an error code. To determine if an error code has been returned, you can use the (vl-catch-all-error-p *arg*) function, where *arg* is the value returned by (vl-catch-all-apply). If (vl-catch-all-error-p *arg*) returns T, then an error has occurred. In this case you can get the error message using the (vl-catch-all-error-message *arg*) function.

Here is a handy function that encapsulates the error handling functions in a single function called **ok**. It takes the same arguments as (vl-catch-all-apply). If an error occurs, the error message is printed, and nil is returned. If no error occurs, then this function returns the same value that *func* would have returned unless it is nil, in which case T is returned.

```
(defun ok(func argList / err)
  (setq err (vl-catch-all-apply func argList))
  (cond
    ((vl-catch-all-error-p err)
      (princ (strcat "\n" (vl-catch-all-error↵
message err)))
      nil
    )
    (err)
    (T)
  );cond
);defun
```

The (ok) function can then be used in place of (vl-catch-all-apply). If an error occurs, the function takes care of notifying the user of the error and simply returns nil. If a non-nil value is returned, then the function was successful. Here is the same example as above, using the (ok) function:

```
(cond

  ; If an error occurs, the user is↵
notified by the ok function
  ((null (ok 'ado-Open
    (list db conString "" ""↵
ado-adOptionUnspecified))))

  ; No error condition. The code resumes here...
  (T
    (princ "\nSuccess!")
```

```
    )

  )
```

RETRIEVING A RECORDSET

In LISP, because it is a list processing language, everything is easier to manage if it is stored in a list. This is especially true when you work with external databases, because a table of rows and columns can easily be represented as a series of nested lists. Shown below is a function that returns a list based on an ADO command (a table name or SQL statement). The function returns a list that has the following structure:

- Each member of the list represents a row in the returned cursor.
- Each row is represented as an association list that contains the data for that row as follows:

```
  ((columnName . value) (columnName . value) ...)

(defun getRecordList(adoConnection↵
strCommand / rsObj fields
  numFields i fieldList rsList rowData)

  (cond

    ; Create a Recordset Object
    ((null (setq rsObj↵
(vlax-create-object "ADODB.Recordset")))
      (princ "\nUnable to create↵
ADODB.Recordset object.")
    )

    ; Open the Recordset
    ((null (ok 'ado-Open (list rsObj strCommand↵
adoConnection
      ado-adOpenDynamic ado-adLockOptimistic ado-↵
adOptionUnspecified))))

    (T
      ; Initialize variables
      (setq
        fields (ado-get-Fields rsObj)
        numFields (ado-get-Count fields)
        i 0
        fieldList nil
        rsList nil
      );setq
```

```
      ; Get a list of field names
      (while (< i numFields)
        (setq
          fieldList (append⤶
fieldList (list (ado-get-Item fields i)))
          i (1+ i)
        );setq
      );while

      ; rsList is where we'll save⤶
the main recordset list
      (setq rsList nil)

      ; Iterate through the recordset
      (while (= (ado-get-EOF rsObj) :vlax-false)

        ; Get the field values
        (setq rowData nil)
        (foreach fieldObj fieldList
          (setq rowData
            (append rowData
              (list
                (cons
                  (ado- get-Name fieldObj)
                  (vlax-⤶
variant-value (ado-get-Value fieldObj))
                );cons
              );list
            );append
          );setq
        );foreach
        (setq rsList (append rsList⤶
(list rowData)))
        (ado-MoveNext rsObj)
      );while
    )
  );cond

  ; Close the recordset object and remove⤶
it from memory
  (if rsObj
    (progn
      (if (= (ado-get-State rsObj)⤶
ado-adStateOpen)
```

```
        (ado-Close rsObj)
      );end if
      (vlax-release-object rsObj)
    );progn
  );end if

  rsList

);defun
```

STORING DRAWING DATA IN A DATABASE

The code that follows is basically an AutoLISP version of the WriteCirclesToDatabase VBA macro that was created earlier in this chapter.

```
(defun WriteCirclesToDatabase( / wsPath↵
conString db circlesRS
                    i ename circleObject)

  ; Get the Data Links path
  (setq wsPath
    (vlax-get-property
      (vlax-get-property
        (vlax-get-property (vlax-get-acad object)
          "Preferences") "Files") "WorkspacePath")
  );setq

  ; Create the connection string
  (setq conString (strcat "File↵
Name=" wsPath "\\circles.udl"))

  (cond

    ; Get the selection set of all↵
circles in the current drawing
    ((null (setq ss (ssget "x"↵
'((0 . "CIRCLE")))))
      (princ "\nNo circles found in this drawing.")
    )

    ; Create the ADO Connection object
    ((null (setq db (vlax-create↵
object "ADODB.Connection")))
      (princ "\nUnable to create↵
ADODB.Connection object.")
    )
```

```
    ; Create the ADO Recordset object
    ((null (setq circlesRS√
(vlax-create-object "ADODB.Recordset")))
      (princ "\nUnable to create↵
ADODB.Recordset object.")
    )

    ; Connect to the database
    ((null (ok 'ado-Open
      (list db conString "" "" ado↵
adOptionUnspecified))))

    ; Delete all rows in the CIRCLES table
    ((null (ok 'ado-Execute
      (list db "DELETE FROM CIRCLES"↵
nil ado adOptionUnspecified))))
    ; Open the circlesRS recordset
    ((null (ok 'ado-Open
      (list circlesRS "CIRCLES" db ado↵
adOpenDynamic
        ado-adLockOptimistic ado↵
adOptionUnspecified))))

    ; Make sure the Recordset supports AddNew
    ((= :vlax-false (ok 'ado-Supports↵
(list circlesRS ado adAddNew)))
      (princ "\nCannot add records to↵
the recordset.")
    )

    (T
      (setq i 0)
      (while (setq ename (ssname ss i))
        ; Convert entity name to object
        (setq circleObject (vlax-ename->vla↵
object ename))

        ; Add a new blank record to↵
the CIRCLES table
        (ado-AddNew circlesRS)

        ; Set the field values
        (setFieldValues circlesRS
          (list
```

```
              (cons "Handle"   ⏎
(vla-get-Handle circleObject))
              (cons "Radius"   ⏎
(vla-get-Radius circleObject))
              (cons "Color"    ⏎
(vla-get-Color  circleObject))
              (cons "Center_X"⏎
(cadr  (assoc 10 (entget ename))))
              (cons "Center_Y"⏎
(cadr (assoc 10 (entget ename))))
            );list
          );setFieldValues

           ; Commit the changes
          (ado-Update circlesRS)

          ; Set index to the next⏎
object in the selection set
          (setq i (1+ i))
        );while
      )
  );cond

  ; Close the recordset object and remove⏎
it from memory
  (if circlesRS
    (progn
      (if (= (ado-get-State circlesRS)⏎
ado adStateOpen)
        (ado-Close circlesRS)
      );end if
      (vlax-release-object circlesRS)
    );progn
  );end if

  ; Close the connection object and remove⏎
it from memory
  (if db
    (progn
      (if (= (ado-get-State db) ado-adStateOpen)
        (ado-Close db)
      );end if
      (vlax-release-object db)
              );progn
```

```
  );end if

  (princ)

);defun

; This function takes a recordset object,⏎
and an assoc list of
; field names and values. A new value is⏎
assigned to each field
; in the current record is according to⏎
valueList
 (defun setFieldValues(rsObject valueList)
  (mapcar
    '(lambda(x)
      (ado-put-Value
        (ado-get-Item (ado-get-Fields⏎
rsObject) (car x))
        (cdr x)
      );ado-put-Value
    );labmda
    valueList
  );mapcar
);defun
```

SUMMARY

In AutoCAD 2000, ADO is the preferred method of talking to databases from within an application. This chapter has provided an overview of ADO and how it can be used in both VBA and Visual LISP. Specifically, this chapter helped you

- Understand the basics of the ADO Object Model
- Be able to use ADO to work with an external database from within an AutoCAD VBA program
- Understand how ADO and other libraries can be accessed from Visual LISP

The next chapter introduces Connectivity Automation Objects (CAO). This object model is used to work with link templates and the links on the objects in your drawing.

REVIEW QUESTIONS

1. What are the two most commonly used objects in the ADO object model?
2. How do you obtain the location of the UDL files in VBA? In Visual LISP?
3. How is a UDL file used to establish a connection with ADO?
4. What function is used in Visual LISP to create new functions based on the methods, properties, and constants defined in a COM library?
5. What error-handling function in Visual LISP can be helpful when you work with COM libraries?

EXERCISE

Implement the two remaining VBA examples found in this chapter using Visual LISP. The VBA macros are ModifyCirclesFromDatabase and ModifyDatabaseFromCircles.

CHAPTER 7

Connectivity Automation Objects (CAO)

OBJECTIVES

After completing this chapter, you will understand how to use the CAO object model to perform the following tasks:

- Get information about link templates
- Create, modify, and delete object links
- Reload the labels
- Handle errors
- Select objects in the drawing based on their link values

INTRODUCTION

The Connectivity Automation Objects (CAO) library is a programming interface that allows you to work with the physical links on AutoCAD objects.

What you *can* do:

- Get information about the link templates in an AutoCAD drawing
- Get information about links on objects
- Create a new link on an object
- Modify an existing link
- Delete a link from an object
- Reload the labels in an AutoCAD drawing

What you *can't* do:

- Create, modify or delete link templates
- Create, modify, delete or inspect label templates
- Create freestanding or attached labels

FILES NEEDED FOR THIS CHAPTER

All of the example code in this chapter is contained in a VBA project file called CAO.dvb, which is included on the CD-ROM. Most of the examples are written to work with parcels3.dwg—the same file that was used in the first two chapters.

THE CAO OBJECT MODEL

Figure 7.1 shows a diagram of the CAO object model. The primary controlling object for CAO is the **DbConnect** object. Your programs will always create at least one instance of the **DbConnect** object. In addition to the **DbConnect** object, the CAO object model consists of three collections: **LinkTemplates**, **Links**, and **Errors**. The collections and objects in the CAO object model are described in detail in the sections that follow.

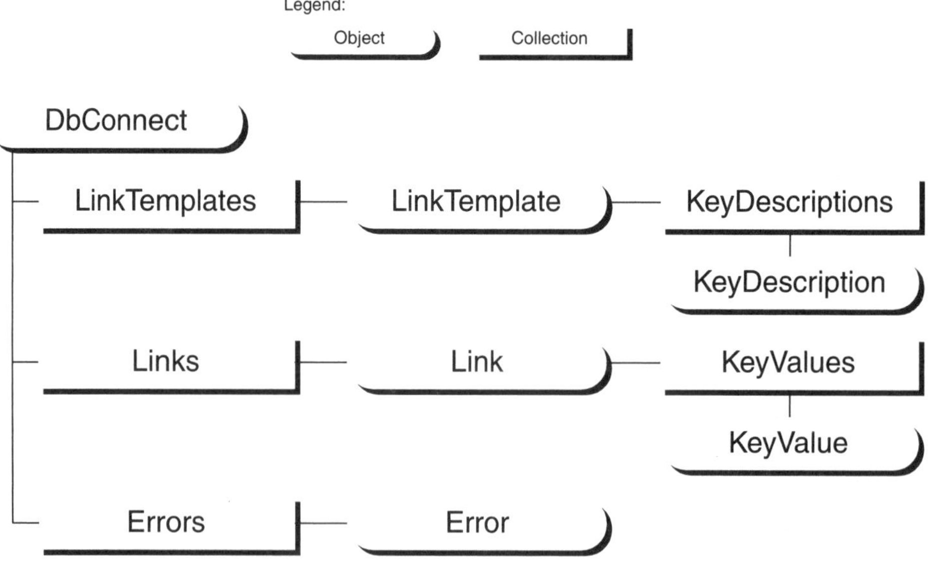

Figure 7.1 *The CAO Object Model*

USING THE CAO LIBRARY IN VBA

Before you can begin using the CAO library in a VBA application, you must reference it in your VBA project. To do this, perform the following steps:

1. From within the VBA editor, select **References** from the **Tools** menu.
2. Locate **CAO 1.0 Type Library** in the list of available references and select the check box next to it.

Note: If you don't see the library in the list, click **Browse** and locate cao15.dll in the directory in which AutoCAD 2000 is installed.

3. Choose **OK**

OVERVIEW OF THE CAO LIBRARY

THE DBCONNECT OBJECT

All of the functionality of the CAO library is accessed through the **DbConnect** object. Before you can do anything with the library, your application must create an instance of the **DbConnect** object. In your VBA application, you can create a **DbConnect** object in the following way:

```
Dim dbc As CAO.DbConnect
Set dbc =↵
AutoCAD.GetInterfaceObject("CAO.DbConnect")
```

The **DbConnect** object has the following methods and properties:

Methods

GetErrors—Gets the **Errors** collection.

GetLinks(*LinkTemplate*, [*ObjectIDs*], [*LinkTypes*])—Gets the links from an object or a group of objects.

GetLinkTemplates([*Document*])—Gets the **LinkTemplates** collection.

ReloadLabels([*ObjectIDs*], [*Document*])—Synchronizes the labels with their associated databases.

Properties

Version—The version number of the CAO library.

DataSourceLocation—The configured directory location for UDL files.

THE LINKTEMPLATES COLLECTION

Information about link templates is obtained through the **LinkTemplates** collection. The **LinkTemplates** collection stores the list of **LinkTemplate** objects for a particular drawing, and it is available through the **GetLinkTemplates** method of the

DbConnect object. Shown below is an example of how to obtain the **LinkTemplates** collection for the current drawing:

```
Dim dbc As CAO.DbConnect
Dim LinkTs As CAO.linkTemplates

Set dbc =↵
AutoCAD.GetInterfaceObject("CAO.DbConnect")
Set LinkTs = dbc.GetLinkTemplates(ThisDrawing)
```

Your code will almost always have some variation of the example above. In order to do just about anything with the CAO library, you must first get the **LinkTemplates** collection for the drawing, and then get the **LinkTemplate** object you want to work with. As you will see, many of the key methods in the CAO object model that allow you to work with links either are members of the **LinkTemplate** object or they require a **LinkTemplate** object as an argument.

The LinkTemplates Collection

The **LinkTemplates** collection has the following methods and properties:

Methods

Item(*Index*)—Retrieves a specific item in the collection by name or index.

Properties

DbConnect—The parent **DbConnect** object.

Document—The parent AutoCAD drawing file.

Count—The number of items in the collection.

To get a specific **LinkTemplate** object, you can use the **Item** method. The **Item** method takes one argument, which can be either an integer or a string. If an integer is supplied, it represents the index (starting with zero) of the **LinkTemplate** object within the collection. If a string is supplied, it represents the name of the link template. For example:

```
Dim LinkTs As CAO.linkTemplates
Dim LinkT As CAO.LinkTemplate
Set LinkT = LinkTs.Item("PARCEL_DATALink1")
```

To iterate through the **LinkTemplates** collection, you can use the **For Each** construct in VBA. Shown below is an example VBA macro that lists the names of the link templates for the current drawing:

```
Sub ListLinkTemplatesExample()
  Dim dbc As CAO.DbConnect
  Dim LinkTs As CAO.linkTemplates
  Dim LinkT As CAO.LinkTemplate
```

```
  ' Create a DbConnect object
  Set dbc =↵
AutoCAD.GetInterfaceObject("CAO.DbConnect")

  ' Get the Link Templates for the current drawing
  Set LinkTs = dbc.GetLinkTemplates(ThisDrawing)

  ' Iterate the LinkTemplates collection and
  ' display the properties
  For Each LinkT In LinkTs
    ThisDrawing.Utility.Prompt vbCrLf & _
      LinkT.Name & ": " & _
      LinkT.DataSource & "." & _
      LinkT.Catalog & "." & _
      LinkT.Schema & "." & _
      LinkT.Table
  Next

  ' Release the DbConnect object
  Set dbc = Nothing
End Sub
```

Running this macro in the parcels3.dwg drawing produces the following output:

```
PARCEL_DATALink1: parcels...PARCEL_DATA
```

The LinkTemplate Object

The **LinkTemplate** object has the following methods and properties:

Methods

CreateLink(*ObjectID, KeyValues*)—Create a new link on an object using the link template.

Properties

DbConnect—The parent **DbConnect** object.

Document—The parent AutoCAD drawing file.

Name—The name given to the link template.

DataSource—The data source (UDL file) name.

Catalog—The catalog name (may not be used for some database systems).

Schema—The schema name (may not be used for some database systems).

Table—The name of the table or view associated with the link template.

KeyDescriptions—A collection of **KeyDescription** objects (described below).

As you learned in the first chapter, a link template stores all the information needed to link multiple objects to a particular table. On the **LinkTemplate** object, the **DataSource**, **Catalog**, **Schema**, **Table**, and **KeyDescriptions** properties provide this information. These properties are read-only, which means you cannot change their values.

The KeyDescriptions Collection

The **KeyDescriptions** property of the **LinkTemplate** object returns a collection of **KeyDescription** objects. A link template can define more than one key column to identify a unique row in the table. Each **KeyDescription** object contains the information about each key column defined for the link template.

The **KeyDescriptions** collection has the following methods and properties:

Methods

Item(*Index*)—Retrieves a specific item in the collection by name or index.

Properties

Count—The number of items in the collection.

DbConnect—The parent **DbConnect** object.

The KeyDescription Object

The **KeyDescription** object has the following properties:

Properties

DbConnect—The parent **DbConnect** object.

FieldName—The name of the link column.

DefinedSize—The maximum number of characters allowed (for string data types).

NumericScale—The maximum number of digits allowed (for numeric data types).

Precision—The maximum number of decimal digits allowed (for numeric data types).

Type—The ADO data type. Possible values are defined by the ADO **DataTypeEnum** enumerated datatype.

Note: Working with data types can be tricky. ADO 2.1 defines approximately 40 different data types. Many of them are similar in type but may have a different name. For example, string or character data could be designated as **adBSTR**, **adChar**, **adWChar**, or **adVarChar**, among others. The OLE DB Provider determines how SQL data types are mapped to ADO data types.

Shown below is an example VBA macro that extends the previous example to list the properties of each KeyDescription object of each link template.

```
Sub KeyDescriptionsExample()
  Dim dbc As CAO.DbConnect
  Dim LinkTs As CAO.linkTemplates
  Dim LinkT As CAO.LinkTemplate
  Dim KeyDesc As CAO.KeyDescription

  ' Create a DbConnect object
  Set dbc =↵
AutoCAD.GetInterfaceObject("CAO.DbConnect")

  ' Get the Link Templates for the current  drawing
  Set LinkTs =  dbc.GetLinkTemplates(ThisDrawing)

  ' Iterate the LinkTemplates collection and
  ' display the properties
  For Each LinkT In LinkTs
    ThisDrawing.Utility.Prompt vbCrLf & _
      LinkT.Name & ": " & _
      LinkT.DataSource & "." & _
      LinkT.Catalog & "." & _
      LinkT.Schema & "." & _
      LinkT.Table
      ' Iterate the KeyDescriptions  collection and
    ' display the properties
    For Each KeyDesc In  LinkT.KeyDescriptions
      ThisDrawing.Utility.Prompt vbCrLf  & " " & _
        KeyDesc.FieldName & ": " & _
        vbCrLf & "  ADO Data Type =↵
"  & KeyDesc.Type & _
        vbCrLf & "  Defined Size ↵
= " & KeyDesc.DefinedSize & _
        vbCrLf & "  Numeric Scale =↵
" & KeyDesc.NumericScale & _
        vbCrLf & "  Precision =  " & KeyDesc.Precision
            Next
        Next

        ' Release DbConnect object
        Set dbc = Nothing
End Sub
```

Running this macro in the parcels3.dwg drawing produces the following output:

```
PARCEL_DATALink1: parcels...PARCEL_DATA
 PARCEL_ID:
  ADO Data Type = 8
  Defined Size  = 13
  Numeric Scale = 0
  Precision  = 0
```

THE LINKS COLLECTION

A **Link** object is obtained through the **Links** collection. A **Links** collection is created through the **GetLinks** method of the **DbConnect** object. Links can be retrieved for a single object, a group of objects, or all the objects in the drawing. However, only the links associated with a single link template can be retrieved at any one time.

The Links Collection

The **Links** collection has the following methods and properties:

Methods

Item(*Index*)—Retrieves a specific item in the collection by name or index.

Properties

Count—The number of items in the collection.

DbConnect—The parent **DbConnect** object.

Document—The parent AutoCAD drawing file.

Shown below is an example VBA macro that extends the ListLinkTemplatesExample() macro to show the number of links that exist for each link template.

```
Sub CountLinksExample()
  Dim dbc As CAO.DbConnect
  Dim LinkTs As CAO.linkTemplates
  Dim LinkT As CAO.LinkTemplate

  ' Create a DbConnect object
  Set dbc =⇙
AutoCAD.GetInterfaceObject("CAO.DbConnect")

  ' Get the Link Templates for the current  drawing
  Set LinkTs =  dbc.GetLinkTemplates(ThisDrawing)

  ' Iterate the LinkTemplates collection and
  ' display the properties
  For Each LinkT In LinkTs
    ThisDrawing.Utility.Prompt vbCrLf & _
```

```
        LinkT.Name & ": " & _
        LinkT.DataSource & "." & _
        LinkT.Catalog & "." & _
        LinkT.Schema & "." & _
        LinkT.Table & "   (" & _
        dbc.GetLinks(LinkT).Count & _
        " Links)"
    Next
    ' Release the DbConnect object
    Set dbc = Nothing
End Sub
```

Running this macro in the parcels3.dwg drawing produces the following output:

```
PARCEL_DATALink1: parcels...PARCEL_DATA  (83 Links)
```

The Link Object

The **Link** object has the following methods and properties:

Methods

Update—Updates a link with new key values.

Delete—Deletes the link.

Properties

DbConnect—The parent **DbConnect** object.

Document—The parent AutoCAD drawing file.

KeyValues—The lookup values for the key column.

LinkTemplate—The link template associated with the link.

LinkType—The type of link. This indicates whether the link is a normal entity link, a freestanding label, or an attached label. Possible values for **LinkType** are

kUnknownLinkType

kEntityLinkType

kFSLabelType

kAttachedLabelType

ObjectId—The AutoCAD-assigned ID of the object that owns the link.

Updatable—A Boolean value indicating whether or not the link can be updated. If the object is on a locked layer, for example, the link would not be updatable.

ERRORS

The Errors Collection

The **Errors** collection has the following methods and properties:

Methods

Item(*Index*)—Retrieves a specific item in the collection by name or index.

Properties

Clear—Removes all the errors from the collection.

Count—The number of items in the collection.

The Error Object

The **Error** object has the following properties:

Properties

ErrorCode—The numeric error code.

ErrorDescription—The text description of the error.

Shown below is an example VBA function that lists the error code and description of any errors that exist in the **Errors** collection.

```
Function showDbErrors(dbc As CAO.DbConnect)
  ' Iterate the Errors collection and display
  ' any error information that exists
  Dim dbcError As CAO.Error
  For Each dbcError In dbc.GetErrors
    ThisDrawing.Utility.Prompt _
      "DbConnect Error " & dbcError.ErrorCode &_
      ": " & dbcError.ErrorDescription & vbCrLf
  Next
End Function
```

PUTTING THE CAO LIBRARY TO WORK

It is important to understand the functional distinction between ADO and CAO. Simply put, ADO is used to communicate with your external database, and CAO is used to work with the AutoCAD linking mechanism. In fact, all of the functionality of the CAO library can be used without an active connection to a data source. Within the context of a real-world application, however, you will likely interweave the use of both libraries in your code. Examples of this are demonstrated in Chapter 8.

GETTING LINK INFORMATION

Retrieving the links from an object or a group of objects is accomplished through the **GetLinks** method of the **DbConnect** object. The syntax for **GetLinks** is as follows:

Set object = dbconnect. **GetLinks(*LinkTemplate, ObjectIDs, LinkTypes*)**

The parameters for the **GetLinks** method are as follows:

LinkTemplate—The **LinkTemplate** object associated with the links you want to retrieve.

ObjectIDs—An array of the AutoCAD object IDs from which the links are retrieved. This argument is optional. If it is omitted, then all links for the link template are retrieved.

LinkTypes—Any combination of link types to search for. This argument is optional. If it is omitted, then only normal entity links are retrieved.

kEntityLinkType—Retrieves normal entity links.

kFSLabelType—Retrieves freestanding label links.

kAttachedLabelType—Retrieves attached label links.

The **GetLinks** method returns a **Links** collection object. If no links are found, then the collection is empty.

The **GetLinks** method requires that you provide a **LinkTemplate** object as an argument. Normally, in the context of an application, you would know which link template you were working with. If you want to get all links for all link templates, you must call **GetLinks** for each link template in the **LinkTemplates** collection and check the **Count** property of the returned **Links** collection.

Writing a Generic Link Retrieval Function

Shown below is a sample VBA macro that prompts the user for an object and then displays information about all links found on that object. For each link found, the link template name is displayed along with the link type value. Then the list of key columns and key values are displayed.

```
Sub GetLinksExample()
  Dim dbc As CAO.DbConnect
  Dim ptPick(0 To 2) As Double
  Dim objEntity As AutoCAD.AcadEntity
  Dim LinkT As CAO.LinkTemplate
  Dim keys As New CAO.KeyValues
  Dim keyval As New CAO.KeyValue
  Dim objLink As CAO.Link
  Dim objLinks As CAO.Links
  Dim idArray(0 To 0) As Long
  Dim objKey As CAO.KeyValue
  Dim linksFound As Boolean
  Dim LinkTypes As Variant

  ' Prompt the user to select an object
  On Error Resume Next
  ThisDrawing.Utility.GetEntity objEntity,↵
ptPick, "Select an object: "
  If Err Then GoTo done
        On Error GoTo 0
```

```
  ' Create a DbConnect object
  Set dbc =↵
AutoCAD.GetInterfaceObject("CAO.DbConnect")

  idArray(0) = objEntity.ObjectID
  LinkTypes = kEntityLinkType Or↵
kFSLabelType Or kFSLabelType
  linksFound = False

  ' Iterate through the link templates  collection
  For Each LinkT In↵
dbc.GetLinkTemplates(ThisDrawing)

    ' Get the links on this object↵
with this link template
    Set objLinks = dbc.GetLinks(LinkT,↵
idArray, LinkTypes)

    ' Any links found?
    If objLinks.Count > 0 Then
      linksFound = True
      ThisDrawing.Utility.Prompt vbCrLf  & "Links↵
for " & _
        LinkT.Name & vbCrLf

      ' Iterate through the collection  of links
      For Each objLink In objLinks
        ThisDrawing.Utility.Prompt  "   Link↵
type: " & _
          objLink.LinkType & vbCrLf

        ' Iterate through the  collection of↵
key values
        For Each objKey In  objLink.KeyValues
          ThisDrawing.Utility.Prompt↵
"    " & objKey.FieldName & _
            " = " & objKey.Value &  vbCrLf
        Next
      Next
    End If
  Next
```

```
  ' If no links were found for any of the↵
link templates
  If Not linksFound Then
    MsgBox "No links found on that object."
  End If

  ' Release the DbConnect object
done:
  Set dbc = Nothing
End Sub
```

Running this macro in the parcels3.dwg drawing and selecting one of the parcel polylines produces the following output:

```
Select an object: <pick a parcel polyline>

Links for PARCEL_DATALink1
  Link type: 2
  PARCEL_ID = 2945-12-42-14
```

CREATING A NEW LINK ON AN OBJECT

Creating a new link on an object is accomplished through the **CreateLink** method of the **LinkTemplate** object. The syntax for **CreateLink** is as follows:

Set object = *linktemplate.CreateLink(ObjectID, Keys)*

The parameters for the **CreateLink** method are as follows:

ObjectID—The ID of the AutoCAD object to which the link is attached.

Keys—A **KeyValues** collection containing the column values for the link.

The **GetLinks** method returns a **Link** object.

Before you can create a link on an object, you must create a new **KeyValues** collection and populate it with **KeyValue** objects. It is only necessary to provide a value for each **KeyValue** object, by setting the **Value** property. It is not necessary to set the **FieldName** property; **CreateLink** ignores it. If your key consists of more than one column, you must supply the key values in exactly the same order as they appear in the table. This is also the same order in which the columns appear in the **KeyDescriptions** collection of the **LinkTemplate** object.

CreateLink Example

The following example VBA function demonstrates the use of the **CreateLink** method. This function takes three arguments: a link template name, an AutoCAD object, and a link value. This function assumes that there is only one link column used.

```
Function CreateLinkExample(strTemplateName↵
As String, _
    objEntity As AutoCAD.AcadObject,  varValue As↵
Variant) As CAO.Link

  On Error Resume Next
  Dim dbc As CAO.DbConnect
  Dim LinkT As CAO.LinkTemplate
  Dim keys As New CAO.KeyValues
  Dim keyval As New CAO.KeyValue

  ' Create a DbConnect object
  Set dbc =↵
AutoCAD.GetInterfaceObject("CAO.DbConnect")

  ' Get the LinkTemplate object
  Set LinkT =  dbc.GetLinkTemplates(ThisDrawing)↵
.Item(strTemplateName)
  If Err <> 0 Then
    MsgBox strTemplateName & "not found."
    GoTo done
  End If

  ' Set the key value and build the↵
KeyValues collection
  keyval.value = varValue
  keys.Add keyval

  ' Create the link
  Set CreateLinkExample =↵
LinkT.CreateLink(objEntity.ObjectID, keys)

done:
  Set dbc = Nothing
End Function
```

The following macro uses the **CreateLinkExample** function to link an AutoCAD object using the **PARCEL_DATALink1** link template found in parcels3.dwg.

```
Sub testCreateLink()
        On Error Resume Next
        Dim ptPick(0 To 2) As Double
        Dim objEntity As AutoCAD.AcadEntity
        Dim objLink As CAO.Link

        ' Prompt the user to select an object
```

```
  ThisDrawing.Utility.GetEntity objEntity,⏎
ptPick, "Select an object: "
  If Not Err Then
    objLink = CreateLinkExample("PARCEL_DATALink1",⏎
objEntity, _"2945-12-42-01")
  End If
End Sub
```

MODIFYING AN EXISTING LINK

Modifying an existing link on an object is accomplished through the **Update** method of the **Link** object. The **Update** method takes no arguments and returns no value. Prior to using the **Update** method, however, you must first get the **Link** object you want to update and then make the desired changes to the values in the **KeyValues** collection.

Update Example

The following example VBA function demonstrates the use of the **Update** method. This function takes three arguments: a link template name, an AutoCAD object, and the new link value. This function assumes that there is only one link column used.

```
Function UpdateLinkExample(strTemplateName As⏎
String, _
    objEntity As AutoCAD.AcadObject,  varValue As⏎
Variant)

  On Error Resume Next
  Dim dbc As CAO.DbConnect
  Dim LinkT As CAO.LinkTemplate
  Dim objLink As CAO.Link
  Dim objLinks As CAO.Links
  Dim idArray(0 To 0) As Long
  Dim linkTypes As Variant

  ' Create a DbConnect object
  Set dbc =⏎
AutoCAD.GetInterfaceObject("CAO.DbConnect")

  ' Get the LinkTemplate object
  Set LinkT = dbc.GetLinkTemplates(ThisDrawing)⏎
.Item(strTemplateName)
  If Err <> 0 Then
    MsgBox strTemplateName & "not found."
    GoTo done
  End If

        ' Get the links collection for the⏎
selected object
```

```
  idArray(0) = objEntity.ObjectID
  linkTypes = kEntityLinkType Or↵
kFSLabelType Or kFSLabelType
  Set objLinks = dbc.GetLinks(LinkT,↵
idArray, linkTypes)
  If Err <> 0 Then
    MsgBox "No links found on that  object."
    GoTo done
  End If

  ' Get the first link on the object
  Set objLink = objLinks.Item(0)

  ' Set the new key value (assumes only↵
one key column)
  objLink.KeyValues.Item(0).value = varValue

  ' Update the link
  objLink.Update

done:
  Set dbc = Nothing
End Function
```

The following macro uses the **UpdateLinkExample** function to update a link using the **PARCEL_DATALink1** link template in parcels3.dwg.

```
Sub testUpdateLink()
  On Error Resume Next
  Dim ptPick(0 To 2) As Double
  Dim objEntity As AutoCAD.AcadEntity

  ' Prompt the user to select an object
  ThisDrawing.Utility.GetEntity objEntity,↵
ptPick, "Select an object: "
  If Not Err Then
    UpdateLinkExample "PARCEL_DATALink1",↵
objEntity, "2945-12-42-63"
  End If
End Sub
```

DELETING A LINK FROM AN OBJECT

Deleting a link from an object is accomplished through the **Delete** method of the **Link** object. The **Delete** method takes no arguments and returns no value. Prior to using the **Delete** method, however, you must first get the **Link** object you want to delete.

Delete Example

The following example VBA function demonstrates the use of the **Delete** method. This function takes two arguments: a link template name and an AutoCAD object. This function deletes all of the links from the object that are associated with the given link template.

```
Function DeleteLinkExample(strTemplateName↵
As String, _
    objEntity As AutoCAD.AcadObject)

  On Error Resume Next
  Dim dbc As CAO.DbConnect
  Dim LinkT As CAO.LinkTemplate
  Dim objLink As CAO.Link
  Dim objLinks As CAO.Links
  Dim idArray(0 To 0) As Long
  Dim linkTypes As Variant

  ' Create a DbConnect object
  Set dbc =↵
AutoCAD.GetInterfaceObject("CAO.DbConnect")

  ' Get the LinkTemplate object
  Set LinkT = dbc.GetLinkTemplates(ThisDrawing)↵
.Item(strTemplateName)
  If Err <> 0 Then
    MsgBox strTemplateName & "not found."
    GoTo done
  End If

  ' Get the links collection for the↵
selected object
  idArray(0) = objEntity.ObjectID
  linkTypes = kEntityLinkType Or↵
kFSLabelType Or kFSLabelType
  Set objLinks = dbc.GetLinks(LinkT,↵
idArray, linkTypes)
  If Err <> 0 Then
    MsgBox "No links found on that  object."
    GoTo done
  End If

  ' Iterate through the Links collection↵
and delete the links
      For Each objLink In objLinks
```

```
    objLink.Delete
  Next

done:
  Set dbc = Nothing
End Function
```

The following macro uses the **DeleteLinkExample** function to delete a link using the **PARCEL_DATALink1** link template in parcels3.dwg.

```
Sub testDeleteLink()
  On Error Resume Next
  Dim ptPick(0 To 2) As Double
  Dim objEntity As AutoCAD.AcadEntity

  ' Prompt the user to select an object
  ThisDrawing.Utility.GetEntity objEntity,↵
ptPick, "Select an object: "
  If Not Err Then
    DeleteLinkExample "PARCEL_DATALink1", objEntity
  End If
End Sub
```

RELOADING LABELS

Reloading the labels in a drawing is accomplished through the **ReloadLabels** method of the **DbConnect** object. The **ReloadLabels** method is the only method in the entire CAO library that actually needs to communicate with the database. However, you do not necessarily need to be connected to any data sources in the dbConnect Manager window. The **ReloadLabels** method will establish the connections itself.

The syntax for **ReloadLabels** is as follows:

dbconnect.**ReloadLabels** [*ObjectID*], [*Document*]

The parameters for the **ReloadLabels** method are as follows:

ObjectID—The AutoCAD object IDs of the labels you want to reload. If omitted, all label objects will be reloaded.

Document—The AutoCAD document for which you want to reload the labels. If omitted, labels are reloaded in the active document.

Shown below is a simple VBA macro that demonstrates the use of the **ReloadLabels** method.

```
Sub ReloadLabelsExample()
  Dim dbc As CAO.DbConnect

  ' Create a DbConnect object
```

```
  Set dbc =⤶
AutoCAD.GetInterfaceObject("CAO.DbConnect")

  ' Reload the labels
  dbc.ReloadLabels

  ' Release the DbConnect object
  Set dbc = Nothing
End Sub
```

CAPTURING ERRORS

When you use the CAO library, the VBA error handler captures most errors. VBA handles its errors using a global **Err** object. You have probably noticed the use of the **Err** object in the previous examples to capture errors returned by CAO methods. For example, in each of the examples, we retrieve the **LinkTemplate** object using the name of the link template. In a condition where the link template does not exist, VBA captures the error information in the **Err** object.

The following code fragment demonstrates this:

```
  ' Get the LinkTemplate object
  Set LinkT = dbc.GetLinkTemplates(ThisDrawing)⤶
.Item(strTemplateName)
  If Err <> 0 Then
    MsgBox strTemplateName & "not found."
    GoTo done
  End If
```

In cases where VBA cannot capture a particular error, CAO provides its own error-capturing mechanism. These errors are available to your application through the **Errors** collection. Each time a CAO method is invoked, the **Errors** collection is cleared. If the method generated any errors that could not otherwise be captured by VBA, they are added to the **Errors** collection.

For example, **ReloadLabels** is a method that may generate errors that are external to VBA. This is because **ReloadLabels** actually communicates with the database, and the database provider might return errors that VBA cannot capture. These types of errors are captured by the **Errors** collection. Below is a revised version of the previous **ReloadLabelsExample** macro that displays any errors that exist in the **Errors** collection.

```
Sub ReloadLabelsExampleWithErrors()
  Dim dbc As CAO.DbConnect
  Dim dbcError As CAO.Error

  ' Create a DbConnect object
```

```
  Set dbc =↵
AutoCAD.GetInterfaceObject("CAO.DbConnect")

  ' Reload the labels
  dbc.ReloadLabels

  ' Display any errors that may have been  generated
  For Each dbcError In dbc.GetErrors
    ThisDrawing.Utility.Prompt "DbConnect  Error " & _
      dbcError.ErrorCode & ": " & _
      dbcError.ErrorDescription & vbCrLf
  Next

  ' Release the DbConnect object
  Set dbc = Nothing
End Sub
```

SELECTING LINKED OBJECTS

Earlier, we learned how to obtain the link value of a selected object. There are times, however, when you want to get an object, or selection of objects, that have a particular link value. Unfortunately, there is no quick method to do this. To accomplish this, you must first get the entire selection of links for a particular link template. Then you must iterate through the collection and search for the desired link value. Once the desired link is found, you can then get the object associated with the link.

Shown below is a VBA function that may come in handy in this situation. This function takes two arguments: the name of the link template and a **KeyValues** object that contains the column values you are searching for. It returns an AutoCAD selection set of the objects that were found.

```
Function selectObjectsByKey(strTemplateName↵
As String, _
    objKeys As CAO.KeyValues) As↵
AutoCAD.AcadSelectionSet

  On Error Resume Next
  Dim dbc As CAO.DbConnect
  Dim LinkT As CAO.LinkTemplate
  Dim objLinks As CAO.Links
  Dim objLink As CAO.Link
  Dim ss As AutoCAD.AcadSelectionSet
  Dim i, j As Integer
  ReDim objArray(0) As AutoCAD.AcadEntity
  Dim keyMatch As Boolean

  ' Get or create the selection set
```

```
  Set ss =  ThisDrawing.SelectionSets↵
.Item("Link_Selection")
  If Err <> 0 Then
    Set ss ThisDrawing.SelectionSets.Add↵
("Link_Selection")
  End If

  ' Clear the selection set
  ss.Clear

  ' Create a DbConnect object
  Set dbc =↵
AutoCAD.GetInterfaceObject("CAO.DbConnect")

  ' Get the LinkTemplate object
  Set LinkT =  dbc.GetLinkTemplates(ThisDrawing)↵
.Item(strTemplateName)
  If Err Then
    MsgBox strTemplateName & "not found."
    GoTo done
  End If

  ' Get the links collection for the link template
and build
  ' an array of AutoCAD objects
  Set objLinks = dbc.GetLinks(LinkT)
  If objLinks.Count > 0 Then
    i = 0
    For Each objLink In objLinks
      keyMatch = True
      For j = 0 To objLink.KeyValues.Count - 1
        If objLink.KeyValues.Item(j).value <>_
          objKeys.Item(j).value Then
        keyMatch = False
      End If
    Next
    If keyMatch Then
      ReDim objArray(0 To i)
      Set objArray(i) = _
        ThisDrawing.ObjectIdToObject↵
(objLink.ObjectID)
      i = i + 1
    End If
        Next
```

```
        ss.AddItems objArray
End If

  ' Set the return value for the function
  Set selectObjectsByKey = ss
  done:
  ' Release the DbConnect object
  Set dbc = Nothing
End Function
```

Shown below is an example macro that demonstrates how the **SelectObjectsByKey** function could be used. This macro searches for objects having a link value of "2945-12-42-01" for the **PARCEL_DATALink1** link template and then highlights them in the drawing.

```
Sub testSelectObjectsByKey()
  Dim keys As New CAO.KeyValues
  Dim keyval As New CAO.KeyValue
  Dim ss As AutoCAD.AcadSelectionSet

  ' Set the key value and build the↵
KeyValues collection
  keyval.value = "2945-12-42-01"
  keys.Add keyval

  Set ss =  selectObjectsByKey("PARCEL_DATALink1",↵
keys)
  ss.Highlight True
End Sub
```

SUMMARY

The CAO object model is critical to an application that needs to work with links on objects. It can be integrated easily with the AutoCAD object model and ADO. This chapter has described each component of the CAO object model and has provided code examples that demonstrate how it is used in an application. Specifically, you have learned how to

- Get information about link templates
- Create, modify, and delete object links
- Reload the labels
- Handle errors
- Select objects in the drawing based on their link values

REVIEW QUESTIONS

1. What are some of the key limitations of the CAO library?
2. Before you can do anything with the CAO library, which object must your application create?
3. To which object does the **GetLinks** method belong?
4. What kinds of errors does the **Errors** collection capture?
5. What is the only method in the CAO library that communicates with the database?

EXERCISES

1. Write a macro that compares the field types returned by an ADO **Recordset** and the CAO **KeyDescriptions** object. What did you find?
2. Write a macro that allows the user to copy a link from one object to another.
3. Create two drawings that have the same link template name but with a different definition. What happens when you insert one in the other?

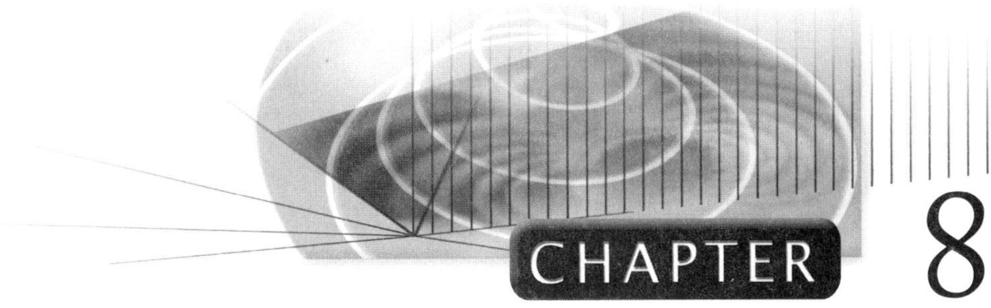

Putting It All Together

OBJECTIVES

The purpose of this chapter is to demonstrate how ADO and CAO can work together in a real-world application. This chapter draws on the knowledge you have gained in all the previous chapters.

INTRODUCTION

To create a robust AutoCAD/Database application, you will likely make use of the CAO and ADO object models, as well as the AutoCAD object model. Chapter 7 focused on using CAO to manage link information in a drawing. This chapter illustrates how to effectively mix all of the programming interfaces in a single working application—specifically, by completing the asset management application that we designed in Chapter 5.

In this chapter, we will be using Visual Basic for Applications (VBA). All the applications that are presented here can also be implemented in Visual LISP. In Chapter 6, we looked at some examples of how the ADO object model can be used in Visual LISP. The concepts are exactly the same for CAO and for any other object model you use in your application.

THE ASSET MANAGEMENT APPLICATION REVISITED

Throughout this chapter, we will be designing and coding the specific macros and applications that were identified as part of the design of our asset management application in Chapter 5. All of the files for the completed application are on the included CD-ROM, in a VBA project file called office.dvb.

Although this application may not match your specific needs exactly, the concepts you learn from the code we develop can easily be applied to your own application. And, of course, we will be focusing on code reuse as well. Many of the functions included with this project are generic enough to be plugged into other applications with little or no modification.

SET UP THE APPLICATION FILES

To start with, our asset management application consists primarily of two files: the floor plan drawing file and the office database. These files are included on the CD-ROM as office.dwg and office.mdb respectively. The office database is a Microsoft Access implementation of the database we designed in Chapter 5. The .mdb file contains only the table structures and relationships; it does not contain any data. Later in this chapter, we will populate the database with the information in the drawing.

We will perform the following steps to set up the environment for this application:

1. Examine the office floor plan drawing.
2. Configure the office data source.
3. Create the link template.

The series of tutorials that follows takes you through this setup process.

Examining the Office Floor Plan Drawing

Our first step is to open and examine the drawing file and make note of the specific characteristics that are important to our application. Specifically, we need to look at the layer structure, the objects that are used to represent things such as spaces and annotation, and what properties exist on those objects. The following tutorial walks you through this process.

TUTORIAL 8.1 – EXAMINING THE DRAWING

1. Launch AutoCAD 2000 and open office.dwg.
2. From the **Format** menu, choose **Layer** to display the **Layer Properties Manager** dialog box, as shown in Figure 8.1.
3. As you familiarize yourself with the layer structure, note the following two layers that will be important to our application (highlighted in Figure 8.1):
   - A-EMPNAM contains the text objects for the room labels
   - A-SPACE contains the closed polylines for the spaces
4. Choose **OK** to exit the **Layer Properties Manager** dialog box.
5. Zoom in on the drawing such that you can see several spaces and read the labels.
6. Make note of the following characteristics of the space polylines and labels:
   - There are closed polylines on the A-SPACE layer that define the perimeter of each space.
   - The polylines for offices are color coded by department.
   - For non-office space, the polyline color is set to "bylayer."

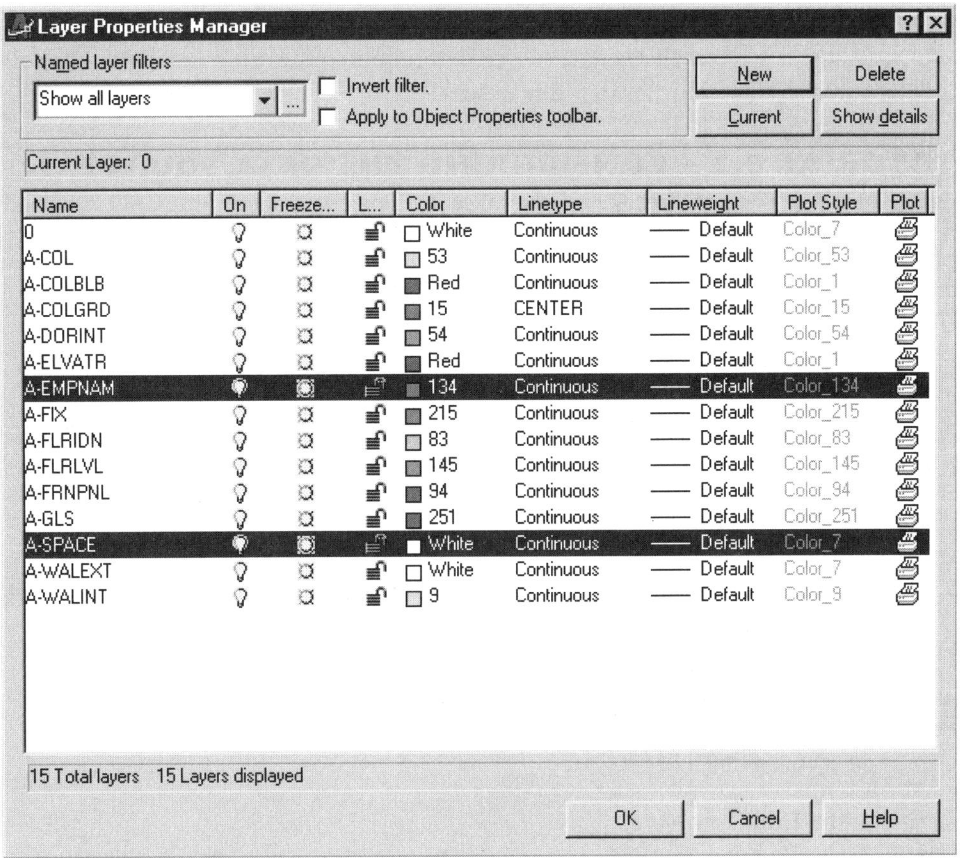

Figure 8.1 *Layer Properties Manager Dialog Box*

- Each space is labeled with a text object, which is on the A-EMPNAM layer.
- For offices and cubicles, the label is the name of the employee and the color is set to "bylayer."
- For vacant offices, the label reads "Vacant" and the text color is set to red.
- For vacant offices, the polyline color is set to "bylayer."
- For non-office space, the label describes the use of the space, and the color is set to yellow.

7. If you are going to proceed immediately with the next tutorial, leave the drawing open. Otherwise, it can be closed.

Configuring the Office Data Source

Our next step is to configure a data source that points to the office database. The following tutorial takes you through this process.

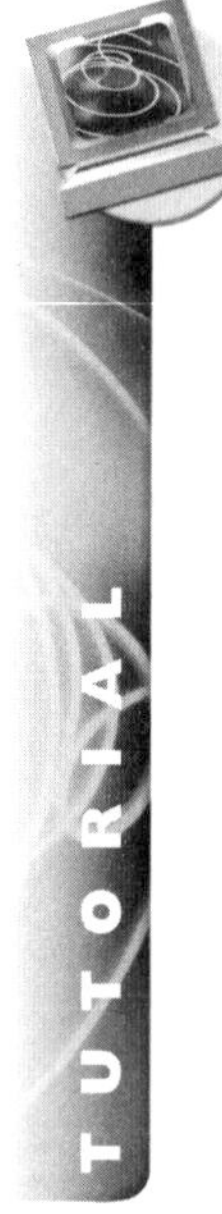

TUTORIAL 8.2 – CONFIGURING THE DATA SOURCE

1. If you do not already have the office drawing open, launch AutoCAD 2000 and open office.dwg.
2. From the **Tools** menu, choose **dbConnect** to display the dbConnect Manager window.
3. Right-click the Data Sources branch in the dbConnect Manager and choose **Configure Data Source...** from the shortcut menu.
4. Type "office" as the data source name, and choose **OK** or press ENTER. This launches the **Data Link Properties** dialog box.
5. Select **Microsoft Jet 3.51 OLE DB Provider** in the list of OLE DB Providers, and choose **Next >>**.

Tip: Double-clicking to select a provider will automatically advance you to the next tab in the dialog box.

6. Enter the full path of the office.mdb file you copied from the CD-ROM.
7. Choose **Test Connection** to verify that a connection can be made.
8. Choose **OK**.

Creating the Link Template

The final step in our setup process is to create the link template that will be used to link the space polylines to the database. At the conclusion of Chapter 5, we decided that the most useful object link in this application is between the space polylines and the **SPACES** table. As we learned in Chapter 4, however, it can give us a great deal of flexibility if we establish our link using a *view* rather than a table. In the office database that was created for this project, there is a view based on the **SPACES** table called **SPACES_QUERY**. This view makes columns from all the other tables available to your application from a single virtual table. Figure 8.2 shows the structure of this view, with all the columns shown.

SPACES_QUERY
SPACE_ID EMPLOYEE_ID LAST_NAME FIRST_NAME TITLE PHONE_EXT DEPT_NAME HAS_WINDOW ENCLOSED TYPE_NAME

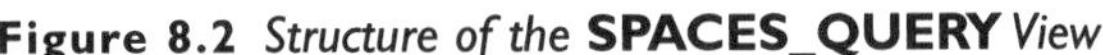

Figure 8.2 *Structure of the* **SPACES_QUERY** *View*

In the following tutorial, we will create a link template using the **SPACES_QUERY** view.

TUTORIAL 8.3 - CREATING THE LINK TEMPLATE

1. If you do not already have the office drawing open, launch AutoCAD 2000 and open office.dwg.
2. Connect to the office data source by double-clicking it in the dbConnect Manager. It will expand to display the available tables and views, as shown in Figure 8.3.
3. Right-click the **SPACES_QUERY** table and choose **New Link Template**. The **New Link Template** dialog box is displayed (Figure 8.4).
4. Accept the default link template name of **SPACES_QUERYLink1** by choosing **Continue**. This takes you to the **Link Template** dialog box.
5. Choose the box next to **SPACE_ID** in the list of key fields and then choose **OK**. You should now see a **SPACES_QUERYLink1** node in the dbConnect Manager just below office.dwg (Figure 8.6).
6. Save the drawing.

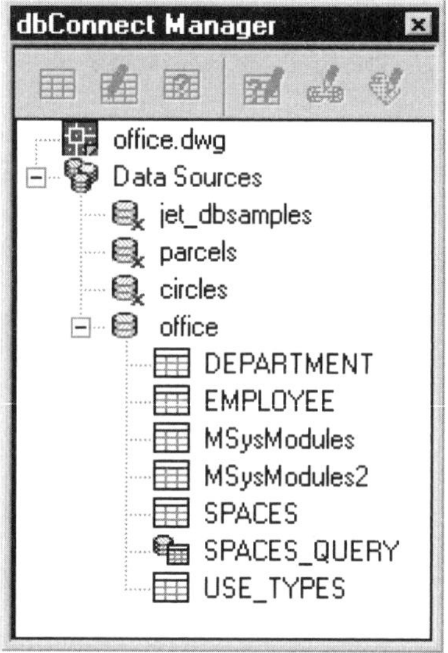

Figure 8.3 *Office Data Source Objects*

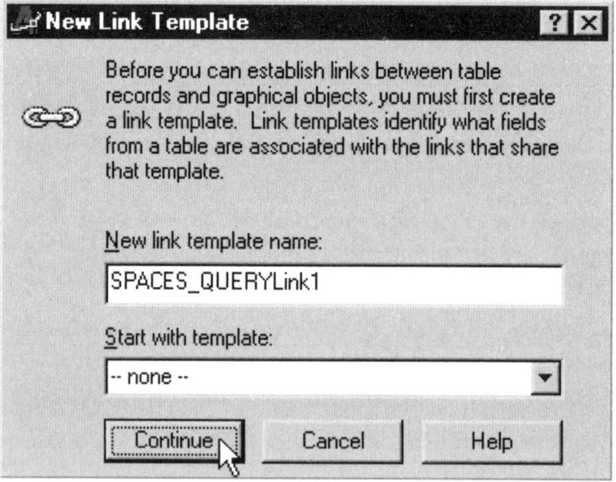

Figure 8.4 *New Link Template Dialog Box*

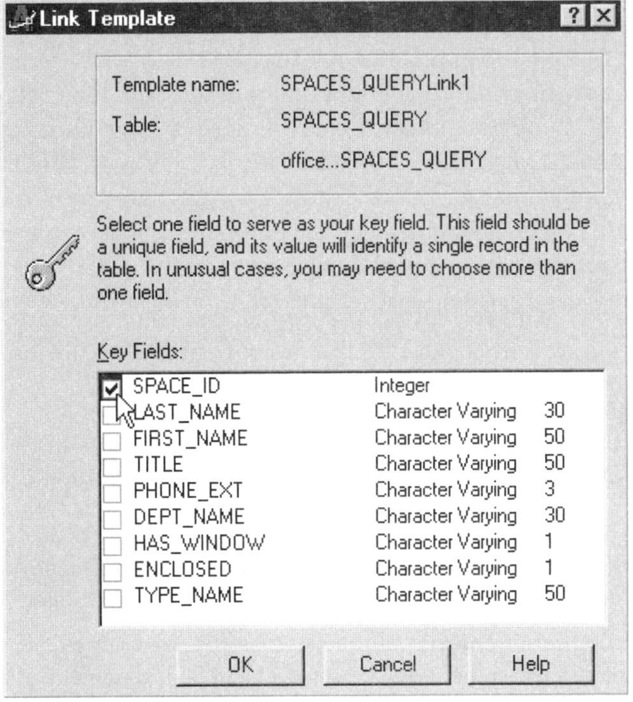

Figure 8.5 *Link Template Dialog Box*

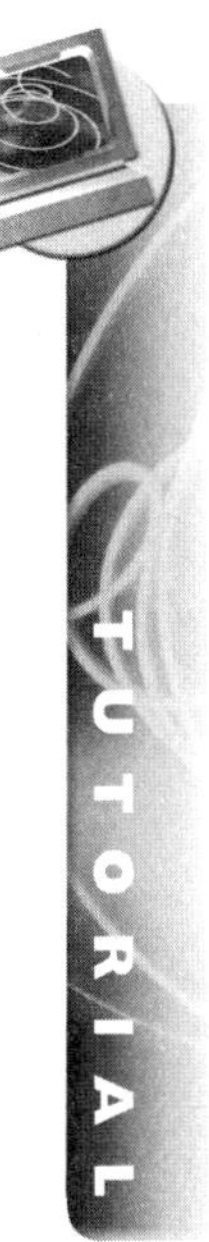

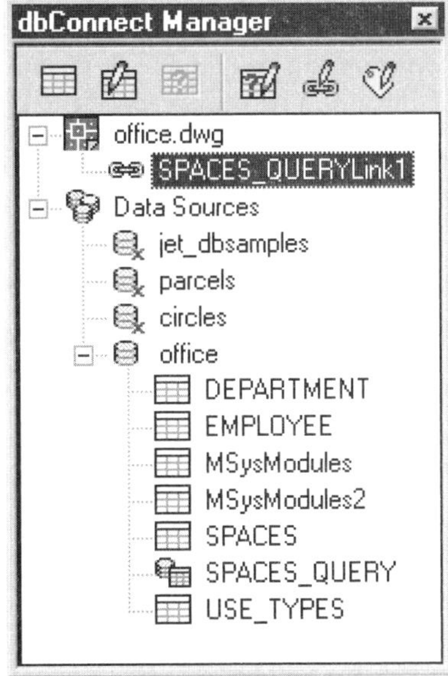

Figure 8.6 *SPACES_QUERYLink1 in the dbConnect Manager*

UNDERSTANDING THE RELATIONSHIP BETWEEN CAO AND ADO

Before you begin writing code, it is important to understand the relationship between the CAO and ADO object models. Once you have a basic understanding of both object models, integrating CAO with ADO is relatively easy. While the two object models are technically independent of one another, there are several properties tucked away in the CAO object model that give the programmer important information that is useful for working with the linked databases through ADO.

Listed below are some of the key properties that can come in handy when ADO is used to communicate with a database that is configured with dbConnect.

dbConnect Properties

DataSourceLocation – Returns the same value as ThisDrawing.Application.Preferences.Files.WorkspacePath.

LinkTemplate Properties

DataSource—This can be used in conjunction with **DataSourceLocation** to connect to the data source with ADO using the UDL file.

Catalog and Schema—If the database you are using supports these components, they can be useful when you build SQL statements that need to specify them.

Table—The table property can be used by your program to create a **Recordset** object when given just a link template.

KeyDescription Properties

FieldName—The field name could be used in your application to build an SQL statement that locates a row for a specific link.

Type—This is the ADO data type. It can be used to help you verify that the value you are assigning to the key is of the correct type, without your having to create a recordset in ADO to get that information.

DefinedSize, NumericScale, and Precision—These properties can be used for data validation as well.

A TYPICAL SCENARIO

To see how this all fits together, let's look at a simple example. One of the most common activities in an AutoCAD/database application is to search a table for the row to which a particular object is linked. To accomplish this, your application performs the following three basic steps:

1. Get the link from the object. This step makes use of AutoCAD objects and CAO. Several examples of this have been provided in Chapter 7.

2. Read the connection information from the associated link template. In this step, we simply gather information from the **LinkTemplate** object.
3. Build the SQL statement to retrieve the desired row. This step uses ADO to create a recordset based on a query built from the information obtained from the **LinkTemplate** object.

As you learned in Chapter 1, link templates store all the common information about the links associated with that link template. Then each individual link simply specifies the link template and the specific key values that identify the unique row in the linked table. Figure 8.7 illustrates the relationships between the AutoCAD object, the link, the link template, and the table. It also shows how the three steps outlined above use the information from each component.

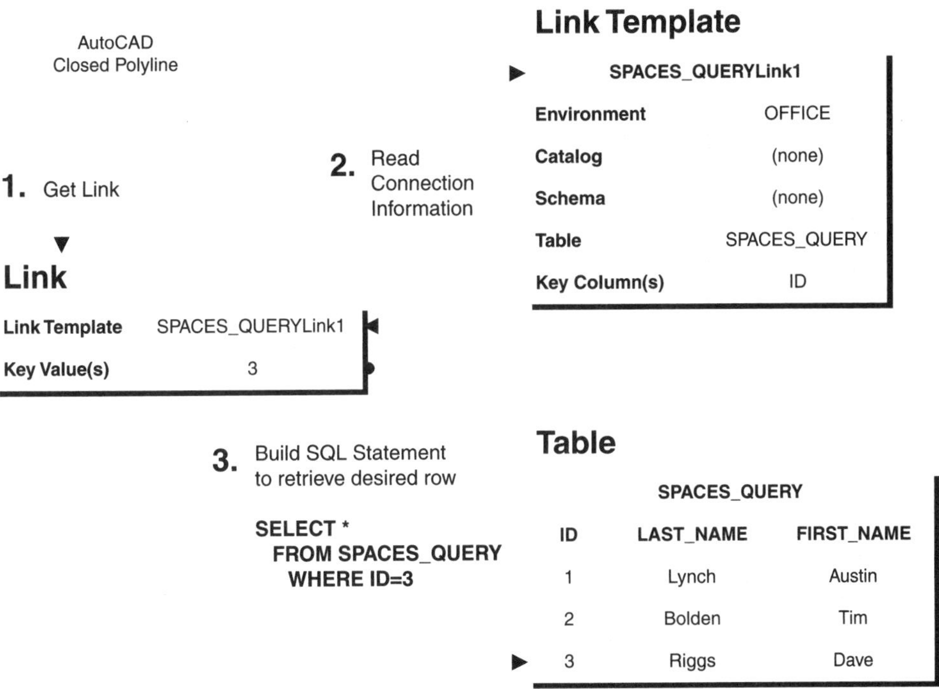

Figure 8.7 *Retrieving a Linked Row from a Table*

So a VBA function that would be useful to us in an application would take an AutoCAD object and a link template name as arguments and return an ADO **Recordset** object. Shown below is an example VBA function that demonstrates how to accomplish this. This function is contained in the module called **modEdit** in the office.dvb VBA project file.

```
Function getRecordsetFromObject(objEntity As↵
AutoCAD.AcadEntity, _
    templateName As String) As ADODB.Recordset

  Dim dbc As CAO.DbConnect
  Dim LinkT As CAO.LinkTemplate
  Dim idArray(0 To 0) As Long
  Dim linkTypes As Variant
  Dim objLink As CAO.link
  Dim linkValue As Variant
  Dim rs As New ADODB.Recordset

  Dim strEnv As String
  Dim strTable As String
  Dim strSQL As String

  ' Create an instance of the DbConnect  Object
  Set dbc =↵
AutoCAD.GetInterfaceObject("CAO.DbConnect")

  ' Get the desired link template from the↵
LinkTemplates collection
  Set LinkT =  dbc.GetLinkTemplates(ThisDrawing)↵
.Item(templateName)

  ' STEP 1
  ' ___________________
  ' Get the links collection from the entity
  ' ___________________
  idArray(0) = objEntity.ObjectID
  linkTypes = CAO.kEntityLinkType Or↵
CAO.kFSLabelType
  On Error Resume Next
  Set objLink = dbc.GetLinks(LinkT, idArray,↵
linkTypes).Item(0)
  If Err Then
    Err.Clear
    MsgBox "That object has no links for "↵
& templateName & "."
```

```
  Exit Function
End If
On Error GoTo 0

' STEP 2
' ------------------
' Read the connection information
' ------------------
strEnv = LinkT.dataSource
strTable = LinkT.Table

' STEP 3
' -------------------------
' Build SQL statement to retrieve desired row
' -------------------------
strConn = "File Name=" &↵
dbc.DataSourceLocation & "\" & strEnv & ".udl"
strSQL = "SELECT * FROM " & strTable & "↵
WHERE " & _
    LinkT.KeyDescriptions.Item(0)↵
.FieldName & "=" & _
    objLink.KeyValues.Item(0).Value

rs.Open strSQL, strConn, adOpenForwardOnly,↵
adLockReadOnly
  Set getRecordsetFromObject = rs
End Function
```

Keep in mind that this function has been written to illustrate the concept of using the information in the link template to create a query on the database. It would take a bit of work to turn it into a generic function. For example, this function makes the following assumptions:

- The data source does not support the catalog and schema.
- The object has only one link for the given link template.
- The link uses a single key column.
- The key column is a numeric data type.

DEVELOPING CAO UTILITY FUNCTIONS

Before we start writing the specific applications for our office project, some utility functions should be written. These functions accomplish basic tasks that are used repeatedly throughout the application. And, more importantly, many of them can be reused in other applications. These functions are contained in the module called **modHelperFunctions** in office.dvb.

The following is a description of each of these utility functions:

getDbConnect()—Returns a global **DbConnect** object that is used throughout the application. Instead of a privately declared **DbConnect** object being created and destroyed within each macro, a single global variable is declared for the application and this function can be used wherever a **DbConnect** object is needed.

getAdoConnection()—Follows the same methodology as **getDbConnect()** except that it returns a global ADO **Connection** object.

getLinkTemplate(*templateName*)—Returns a **LinkTemplate** object from a link template name. This function encapsulates all of the error handling for this simple procedure. If an error occurs, the appropriate error message is displayed, and the function returns **Nothing**.

getConnectionString(*dataSource*)—Returns the connection string for the given data source. It simply uses the "File Name" form of the connection string with the path to the UDL file.

openConnection(***templateName***)—Opens a connection to the data source associated with the given link template name using the global ADO **Connection** object. It returns **True** if successful, or **False** if not.

openRecordset(***templateName***)—Returns an ADO **Recordset** object based on the table or view associated with the given link template name.

getLinkValueFromEntity(***objEntity***, ***templateName***)—Returns the value of the link on the given entity, which is associated with the given link template name.

createLinkOnEntity(***objEntity***, ***templateName***, ***keyValue***)—Creates a new link on the given object using the given key value. Assumes that only one link column is used, and only one link will exist on the entity. Returns the new **Link** object.

This module also contains the following functions that are used for handling the closed polygons:

getSpacePolylineSelection()—Returns an **AcadSelectionSet** object containing all of the space polylines.

SelectPolylinesOnScreen()—Prompts the user for a selection of space polylines.

selectPolylineFromPoint()—Prompts the user to select a point and returns the space polyline (AcadLWPolyline) in which that point resides. This function provides the user a more intuitive way to select a space than having to pick the actual polyline object, which overlaps other lines on the drawing.

getPolylineFromPoint(*point*)—Called by the **selectPolylineFromPoint()** function. It is also used to determine to which polyline a particular label belongs. It also returns a space polyline object as an AcadLWPolyline.

IsInside(*checkPoint, objPline*)—Returns **True** if the given point is inside the given polyline. It is used by **selectPolylineFromPoint()** and **getPolylineFromPoint()**.

getCenterOf(*objPline*)—Returns a point that represents the center of the polyline's bounding box.

getTextInsidePolygon(*objPline*)—Returns the first text object found inside the given polyline object.

In addition to these functions, the following global variables are declared in **modHelper**Functions:

```
Public Const SpacesLink = "SPACES_QUERYLink1"
Public Const SpacesLayer = "A-SPACE"
Public Const TextLayer = "A-EMPNAM"
Public dbc As CAO.DbConnect
Public db As ADODB.Connection
```

DEVELOPING THE APPLICATIONS

The next task is to start developing the actual applications that will allow users to interact with the database using the AutoCAD drawing. The applications are divided into the following categories:

- Data Creation
- Data Maintenance
- Data Integrity Validation
- Query and Annotation

DATA CREATION APPLICATIONS

Populating the Database and Linking the Polylines

While our office floor plan drawing may have seemed simple at first, our examination revealed that it actually has a tremendous amount of useful information in it. In fact, we can just about populate our entire database using this information. And while we're at it, we can create the links on the space polylines. The best part is that we can write an application to do the work for us!

What we need is a simple VBA program that will scan the drawing and populate the tables in our database. The basic flow of this program is as follows:

1. Get a selection set of all the space polylines.
2. For each polyline, find the text that is enclosed within it.

3. Determine the use of the space. If the polyline color is set, then it's an office and the text is the employee name. Otherwise the text is the use type.
4. Add the use type to the **USE_TYPES** table if it does not exist.
5. Add a row to the **SPACES** table for the polyline and create the link on the polyline.
6. If the space is an office, determine the department of the polyline and add the department name to the **DEPARTMENT** table if necessary.
7. Add the employee name to the **EMPLOYEE** table.

A critical piece of information we do not have on the drawing is which polyline color corresponds to which department name. We will assume that we retrieved this information from the facilities manager. Table 8.1 shows the color/department relationship:

Note: Since populating the database only needs to happen once, we will hard-code this list into the macro.

Color	Department Name
1	Civil Engineering
2	Architecture
3	Planning
4	Landscape Architecture
5	Surveying
6	Admin

TABLE 8.1 *Department Colors for Office Drawing*

The following VBA macro accomplishes the task of populating the database and linking the space polylines.

```
Sub populateOfficeDatabase()

  ' Variables for AutoCAD objects
  Dim objPoly As AutoCAD.AcadLWPolyline
  Dim objText As AutoCAD.AcadText
  Dim polySel As AutoCAD.AcadSelectionSet

  ' Variables for ADO objects
  Dim rsSpaces As New ADODB.Recordset
  Dim rsUseTypes As New ADODB.Recordset
```

```
  Dim rsEmployee As New ADODB.Recordset
  Dim rsDepartment As New ADODB.Recordset

  ' Variables for DCO objects
  Dim LinkT As CAO.LinkTemplate
  Dim keys As New CAO.KeyValues
  Dim keyval As New CAO.KeyValue
  Dim objLink As CAO.link

  ' Other variables
  Dim wsPath As String
  Dim useType As String
  Dim DeptName(1 To 6) As String
  Dim i As Long

  ' Store the department names in an array
  DeptName(1) = "Civil Engineering"
  DeptName(2) = "Architecture"
  DeptName(3) = "Planning"
  DeptName(4) = "Landscape Architecture"
  DeptName(5) = "Surveying"
  DeptName(6) = "Admin"

  ' Verify that the SPACES link template  exists
  On Error Resume Next
  Set LinkT = getLinkTemplate(SpacesLink)
  If Err <> 0 Then
    MsgBox "You must first create a Link↩
Template named '" & _
      SpacesLink & "'" & _
      vbCrLf & "based on the↩
SPACES_QUERY table (SPACE_ID field)", _
      vbOKOnly, "Office Example"
    Exit Sub
  End If
  On Error GoTo 0

  ' Verify that the user really wants to  do this
  If MsgBox("This will clear the OFFICE↩
database " & _
      "and delete all links. Do you↩
want  to contiue?", _
      vbYesNo, "Office Example") =  vbNo Then
    Exit Sub
  End If
```

```
  ' Delete all links associated with this↵
link template
  ThisDrawing.Utility.Prompt "Deleting↵
links..." & vbCrLf
  For Each objLink In↵
getDbConnect().GetLinks(LinkT)
    objLink.Delete
  Next

  ' Open the database
  openConnection SpacesLink

  ' Delete all rows from all tables
  ' (in case we run this macro more than  once)
  ThisDrawing.Utility.Prompt "Clearing↵
the database..." & vbCrLf
  getAdoConnection().Execute "delete  from employee"
  getAdoConnection().Execute "delete  from spaces"
  getAdoConnection().Execute "delete  from use_types"
  getAdoConnection().Execute "delete↵
from department"

  ' Open the Recordset objects
  rsSpaces.Open "SPACES",  getAdoConnection(), _
    adOpenDynamic, adLockOptimistic
  rsUseTypes.Open "USE_TYPES",  getAdoConnection(), _
    adOpenDynamic, adLockOptimistic
  rsEmployee.Open "EMPLOYEE",  getAdoConnection(), _
    adOpenDynamic, adLockOptimistic
  rsDepartment.Open "DEPARTMENT",↵
getAdoConnection(), _
    adOpenDynamic, adLockOptimistic

  ' Iterate through the space polygons↵
and populate the database
  Set polySel = getSpacePolylineSelection()
  For Each objPoly In polySel

    ' Find a text object inside the  polygon
    Set objText =  getTextInsidePolygon(objPoly)
    If Not objText Is Nothing Then

      ' Determine what the use of↵
the space is. If the polygon
```

```
      ' color is set, then it's an↵
office and the text is the
      ' employee name or "Vacant".↵
Otherwise use the text value
      ' found inside the polygon as  the use type.
      useType = "Office"
      If objPoly.Color = acByLayer Then
        useType = objText.TextString
        If useType = "Vacant" Then
          useType = "Office"
        End If
      End If

      ' Add the use type to the↵
USE_TYPES table if it does not exist
      rsUseTypes.Find "TYPE_NAME='"  & useType & _
        "'", , , adBookmarkFirst
      If rsUseTypes.EOF Then
        rsUseTypes.AddNew
        rsUseTypes!TYPE_NAME = useType
        rsUseTypes.Update
      End If

      ' Add a new row to the spaces  table
      rsSpaces.AddNew
      rsSpaces!TYPE_ID =  rsUseTypes!TYPE_ID
      rsSpaces.Update

      ' Set up the key value for the  link
      keys.Clear
      keyval.Value = rsSpaces!SPACE_ID
      keys.Add keyval

      ' Create the link on the polygon
      Set objLink =↵
LinkT.CreateLink(objPoly. ObjectID, keys)
      ' This is an occupied office...
      If objText.Color = acByLayer _
        And objPoly.Color <> acBy  Layer Then

        ' Add the department name to  the DEPARTMENT
        ' table if necessary
        rsDepartment.Find  "DEPT_NAME='" & _
          DeptName(objPoly.Color)↵
& "'", , , adBookmarkFirst
```

```
        If rsDepartment.EOF Then
          rsDepartment.AddNew
          rsDepartment!DEPT_NAME↵
= DeptName (objPoly.Color)
          rsDepartment.Update
        End If

        ' Split the first and last↵
name by finding the position' of the  separating space
        i = InStr(1,↵
objText.TextString, " ", vbTextCompare)

        ' Add a new row to the↵
employee table and populate it
        rsEmployee.AddNew
        rsEmployee!FIRST_NAME↵
= Left(objText.TextString, i - 1)
        rsEmployee!LAST_NAME =↵
Mid(objText.TextString, i + 1)
        rsEmployee!DEPT_ID =  rsDepartment!DEPT_ID
        rsEmployee!SPACE_ID =  rsSpaces!SPACE_ID
        rsEmployee.Update
      End If
      ThisDrawing.Utility.Prompt  objText.TextString
    End If
  Next

  ' Close the recordset objects
  rsSpaces.Close
  rsUseTypes.Close
  rsEmployee.Close
  rsDepartment.Close

  ' Release the polyline selection object
  Set polySel = Nothing
End Sub
```

Populating Other Columns

In the **SPACES** table, there are two columns that cannot be populated automatically from the drawing. They are **enclosed** and **has_window**. Two macros will be necessary to help the user populate these columns using graphical selection.

Since the two macros will be very similar, we will first create a function that can be used by both macros called **SetSpacesYesNoField**. This function takes two arguments: the name of the yes/no field and a prompt for the user. The **getLinkValueFromSpace** helper function (described earlier) is used to get the link value from each space. Then, rather than using a Recordset object, we simply compose an **UPDATE** statement and send it to the database using the **Execute** method of the **Connection** object.

```
Function SetSpacesYesNoField(strFieldName As↵
String, strPrompt As String)
  ' Declare CAO objects
  Dim dbc As CAO.DbConnect
  Dim LinkT As CAO.LinkTemplate

  ' Declare AutoCAD objects
  Dim ssPoly As AutoCAD.AcadSelectionSet
  Dim objPoly As AutoCAD.AcadLWPolyline

  Dim spaceID As Long
  Dim strYesNo As String

  ' Prompt the user to select some space↵
polyline objects
  Set ssPoly = SelectPolylinesOnScreen()

  ' Does the selection set contain any  objects?
  If ssPoly Is Nothing Then Exit Function
  If ssPoly.Count = 0 Then Exit Function

  ThisDrawing.Utility.InitializeUserInput  1, "Y N"
  strYesNo = ThisDrawing.Utility.GetKeyword  _
    ("Are these spaces " & strPrompt &  " [Yes/No]? ")

  ' Open the database connection
  openConnection SpacesLink

  ' Iterate through the selection and set  has_window
  ' to the appropriate value
  For Each objPoly In ssPoly
    spaceID =  getLinkValueFromSpace(objPoly)
    getAdoConnection().Execute "update↵
spaces set " & strFieldName & _
    "='" & strYesNo & "'" &  " where↵
space_id=" & Str(spaceID)
```

```
    Next

    ' Close the connection
    getAdoConnection().Close
End Function
```

Once this function has been defined, creating the two functions, **SetHasWindow** and **SetEnclosed**, is easy.

```
Sub SetHasWindow()
    SetSpacesYesNoField "has_window",↵
"adjacent to a window"
End Sub

Sub SetEnclosed()
    SetSpacesYesNoField "enclosed", "enclosed"
End Sub
```

DATA MAINTENANCE APPLICATIONS

Several data maintenance applications were identified in Chapter 5, such as adding new employees, removing employees, and moving employees. For the purposes of demonstration, we will implement an employee removal macro.

Removing an Employee

The following macro prompts the user to select a space (by selecting a point inside a polyline). Then, if the space is an occupied office, the user has the option to remove the employee from the database and convert the space to a "Vacant" space.

```
Sub removeEmployee()
    Dim objPoly As AutoCAD.AcadLWPolyline
    Dim objText As AutoCAD.AcadText
    Dim rsSpaces As ADODB.Recordset
    Dim spaceID As Long

    ' Prompt the user to select a polyline
    Set objPoly = selectPolylineFromPoint()
    If objPoly Is Nothing Then Exit Sub

    ' Open the recordset
    Set rsSpaces = openRecordset(SpacesLink)

    ' Get the link value
    spaceID = getLinkValueFromSpace(objPoly)

    ' Find the corresponding row
```

```
  rsSpaces.Find "SPACE_ID=" & Str(spaceID),⏎
, , adBookmarkFirst

  ' Was the row found?
  If Not rsSpaces.EOF Then
    Select Case rsSpaces!TYPE_NAME

      ' Is this an Office?
      Case "Office"

        ' Is the space already vacant?
        If IsNull(rsSpaces!FIRST_NAME) Then
          MsgBox "That space is already vacant."

        ' Verify that the user wants to do this
        ElseIf MsgBox("Are you sure⏎
you want to remove " & _
          rsSpaces!FIRST_NAME & "⏎
" & rsSpaces!LAST_NAME & _
          "?", vbYesNo, "Remove⏎
Employee") = vbYes Then

          ' Delete the employee from the table
          db.Execute "delete from⏎
employee where " & _
            "space_id=" & Str(spaceID)
          ' Change the space to a "Vacant" space
          objPoly.Color = acByLayer
          Set objText = getTextInsidePolygon(objPoly)
          If Not objText Is Nothing⏎
Then
            objText.TextString = "Vacant"
            objText.Color = 1
          End If
        End If

      Case Else
        MsgBox "That space is not an office."
    End Select
  Else
    MsgBox "Space ID " & Str(spaceID) & "⏎
not found in database."
  End If
```

```
  ' Close the recordset
  rsSpaces.Close
End Sub
```

DATA INTEGRITY VALIDATION APPLICATIONS

In Chapter 5, we discussed some of the issues we face when integrating a CAD environment with a database system. Many of the problems will need to be addressed by our application, such as maintaining the integrity of the one-to-one link between the space polylines and the **SPACES** table.

The AnnotateSpaces and populateOfficeDatabase macros have been written in such a way that at any time, they can be executed to either update the drawing according to the database or update the database according to the drawing. This gives us one level of data integrity validation.

Another possible problem that may occur is that an AutoCAD user might make a copy of a space polyline, thus duplicating the link. The following macro, although not quite complete, demonstrates how to check for duplicate links. It is incomplete because it only identifies the *number* of duplicate links and does not show the user where the duplicates are in the drawing. The function serves its purpose in demonstrating the concept.

```
Sub CheckSpacesLinks()
  Dim ltSpaces As CAO.LinkTemplate
  Dim objLink As CAO.link
  Dim objLinks As CAO.Links
  Dim keys(0 To 0) As Long
  Dim i, j, n, dups As Integer

  ' Get the spaces link template
  Set ltSpaces = getLinkTemplate(SpacesLink)

  ' Get the Links collection
  Set objLinks =↵
ltSpaces.DbConnect.GetLinks(ltSpaces)

  ReDim Key(0 To objLinks.Count)
  i = 0
  n = 0
  dups = 0

  ' Iterate through the Links collection
  For Each objLink In objLinks
    Key(i) =  objLink.KeyValues.Item(0).Value
    If i > 0 Then
```

```
        For j = 0 To i - 1
          If Key(i) = Key(j) Then
            dups = dups + 1
          End If
        Next
      End If
      i = i + 1
    Next
    ThisDrawing.Utility.Prompt vbCrLf & Str(dups) & _
      " duplicate links found."
  End Sub
```

QUERY AND ANNOTATION APPLICATIONS

Annotating the Spaces

In Chapter 5 we identified the need for an application that automatically annotates the drawing according to our own specifications. If the space is an office, then we want to show the employee name unless the office is vacant, in which case we want to show the word "Vacant." For non-office space, we want to show the type of space. Shown below is a macro that accomplishes this.

```
Sub AnnotateSpaces()
  Dim ltSpaces As CAO.LinkTemplate
  Dim objLink As CAO.link
  Dim objPoly As AutoCAD.AcadLWPolyline
  Dim newMtext As AutoCAD.AcadText
  Dim rsSpaces As New ADODB.Recordset
  Dim strKeyField As String

  ' Get the spaces link template
  Set ltSpaces = getLinkTemplate(SpacesLink)
  strKeyField =↵
ltSpaces.KeyDescriptions.Item(0).FieldName & "="

  ' Open the database connection
  openConnection SpacesLink

  ' Open the recordset
  rsSpaces.Open ltSpaces.Table, getAdoConnection(), _
    adOpenStatic, adLockReadOnly

  ' Delete all the annotation
  deleteAnnotation

  ' Iterate through the Links collection
```

```
  For Each objLink In↵
ltSpaces.DbConnect.GetLinks(ltSpaces)
    rsSpaces.Find strKeyField &↵
objLink.KeyValues.Item(0).Value _
      , , , adBookmarkFirst
    If Not rsSpaces.EOF Then
      Set objPoly =  ThisDrawing.ObjectIdToObject↵
(objLink.ObjectID)
      If rsSpaces!TYPE_NAME = "Office"  Then
        If IsNull(rsSpaces!FIRST_NAME)  Then
          Set newMtext = _
            CreateTextObject(getCenterOf(objPoly),↵
"Vacant")
          newMtext.Color = 1
        Else
          Set newMtext = _
            CreateTextObject(getCenterOf(objPoly), _
            rsSpaces!FIRST_NAME↵
& " " & rsSpaces!LAST_NAME)
        End If
      Else
        Set newMtext = _
CreateTextObject(getCenterOf(objPoly), _
            rsSpaces!TYPE_NAME)
        newMtext.Color = 2
      End If
    End If
  Next
  rsSpaces.Close
End Sub
```

The AnnotateSpaces macro uses the following function to create the text objects:

```
Function CreateTextObject(insPt As Variant,↵
text As String) _
    As AutoCAD.AcadText
  Set CreateTextObject = _
    ThisDrawing.ModelSpace.AddText(text,  insPt, 6#)
  CreateTextObject.Alignment =↵
acAlignmentMiddleCenter
  CreateTextObject.TextAlignmentPoint = insPt
  CreateTextObject.Layer = TextLayer
End Function
```

The AnnotateSpaces macro uses the following function to delete the text objects:

```
Function deleteAnnotation()
  Dim objSel As AutoCAD.AcadSelectionSet
  Dim objEntity As AutoCAD.AcadEntity
  Dim filterType(0) As Integer
  Dim filterData(0) As Variant

  ' Get or create the selection set
  On Error Resume Next
  Set objSel =  ThisDrawing.SelectionSets↵
.Item("TMP_Selection")
  If Err Then
    Set objSel = ThisDrawing.SelectionSets.Add↵
("TMP_Selection")
  End If
  On Error GoTo 0

  ' Set up selection Filter
  filterType(0) = 8
  filterData(0) = TextLayer

  ' Select the objects
  objSel.Clear
  objSel.Select acSelectionSetAll, , ,↵
filterType, filterData

  ' Delete all object in selection set
  For Each objEntity In objSel
    objEntity.Delete
  Next

  ' Delete the selection set
  objSel.Delete
End Function
```

Hatching Spaces

Since the spaces are represented with closed polylines, creating a color-coded floor plan is quite easy. First, we will create a function that takes a polyline object and hatch color as arguments and hatches the polyline. Then we can write other routines that color code the floor plan by some set of criteria.

Creating the Hatch Function

The following macro creates a single hatch object:

```
Function createHatch(poly As↩
AutoCAD.AcadLWPolyline, hColor As Integer)
  Dim objHatch As AutoCAD.AcadHatch
  Dim outerLoop(0) As AcadEntity
  Dim patternName As String
  Dim PatternType As Long
  Dim bAssociativity As Boolean

  ' Define the hatch
  patternName = "ANSI31"
  PatternType = 0
  bAssociativity = True

  ' Create the associative Hatch object
  Set objHatch =↩
ThisDrawing.ModelSpace.AddHatch(_
    PatternType, patternName,  bAssociativity)
  objHatch.PatternScale = 50
  objHatch.Color = hColor

  ' Create the outer loop for the hatch.
  Set outerLoop(0) = poly

  ' Append the outer loop to the hatch↩
object, and display the hatch
  objHatch.AppendOuterLoop outerLoop
  objHatch.Evaluate
End Function
```

Hatching Spaces by Department

Since the polylines are already color coded by department, we can create hatch objects with the department color simply by reading the color of the polyline. There is no need to access the database.

```
Sub showHatch()
  Dim objPoly As AutoCAD.AcadLWPolyline
  Dim polySel As AutoCAD.AcadSelectionSet
  For Each objPoly In  getSpacePolylineSelection()
    createHatch objPoly, objPoly.Color
  Next
End Sub
```

Hatching Vacant Space

Here is a more complex example, which finds the linked row for each space and hatches the polyline in a red color if the space is tagged as "Vacant." All other space polylines are ignored.

```
Sub hatchVacantSpaces()
  Dim ltSpaces As CAO.LinkTemplate
  Dim objLink As CAO.link
  Dim objPoly As AutoCAD.AcadLWPolyline
  Dim rsSpaces As New ADODB.Recordset
  Dim strKeyField As String

  ' Get the spaces link template
  Set ltSpaces = getLinkTemplate(SpacesLink)
  strKeyField =  ltSpaces.KeyDescriptions.Item(0)↵
.FieldName & "="

  ' Open the database connection
  openConnection SpacesLink

  ' Open the recordset
  rsSpaces.Open ltSpaces.Table,  getAdoConnection(), _
    adOpenStatic, adLockReadOnly

  ' Iterate through the Links collection↵
and hatch vacant spaces
  For Each objLink In↵
ltSpaces.DbConnect.GetLinks(ltSpaces)
    rsSpaces.Find strKeyField &↵
objLink.KeyValues. Item(0).Value _
      , , , adBookmarkFirst
    If Not rsSpaces.EOF Then
      If rsSpaces!TYPE_NAME = "Office"  Then
        If IsNull(rsSpaces!FIRST_NAME)  Then
          Set objPoly = _
          ThisDrawing.↵
ObjectIdToObject (objLink.ObjectID)
          createHatch objPoly, 1
        End If
      End If
    End If
  Next
  rsSpaces.Close
End Sub
```

Deleting Hatching

This macro will clean things up after a hatching routine has been run.

```
Sub deleteHatch()
  Dim hatchSel As AutoCAD.AcadSelectionSet
  Dim objHatch As AutoCAD.AcadHatch
  Dim filterType(0) As Integer
  Dim filterData(0) As Variant

  ' Get or create the selection set
  On Error Resume Next
  Set hatchSel =  ThisDrawing.SelectionSets↵
.Item("Hatch_Selection")
  If Err <> 0 Then
    Set hatchSel =  ThisDrawing.SelectionSets.Add↵
("Hatch_Selection")
  End If
  On Error GoTo 0

  ' Set up selection Filter
  filterType(0) = 0
  filterData(0) = "HATCH"

  ' Select all HATCH objects
  hatchSel.Clear
  hatchSel.Select acSelectionSetAll, , ,↵
filterType, filterData

  ' Delete all object in selection set
  For Each objHatch In hatchSel
    objHatch.Delete
  Next

  ' Delete the selection set
  hatchSel.Delete
End Sub
```

SUMMARY

This chapter has presented several concrete examples of using VBA with the ADO and CAO libraries in a working application. There is plenty of additional functionality that could be incorporated into this application. The possibilities are no longer limited by your knowledge; they are limited only by your imagination. You now have a good foundation upon which you can begin developing your own robust AutoCAD/database applications.

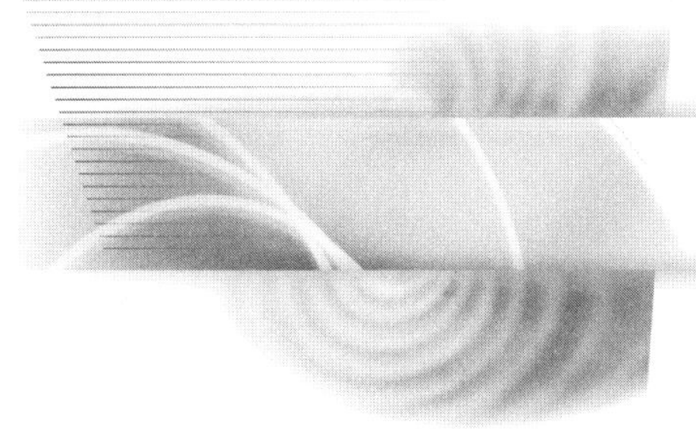

GLOSSARY

ActiveX Data Objects (ADO)
An object model that provides an easy-to-use set of objects, methods, and properties for accessing data in databases. ADO is the COM programming interface to OLE DB.

Aggregate Function
A function in SQL, such as **COUNT**, **SUM**, and **AVG**, used in creating a query that calculates totals or summarizes data.

Alias
An alternate name given to a column, expression, or table in an SQL **SELECT** statement, often shorter or more meaningful.

Application Programming Interface (API)
A feature of a software application that allows users to create custom user interfaces and applications. AutoCAD's APIs include Visual LISP, VBA, and ObjectARX.

Asynchronous Execution
A mode of operation in which a process, such as a query, is initiated by a program, and execution of the program continues before the process is completed. An event is used to notify the application when the process is completed.

Attached Label
See Label.

Boolean Operator
A logical operator used to combine two query expressions. Boolean operators in SQL are **AND**, **OR**, and **NOT**.

Catalog
A level in the database hierarchy representing a database that is available within an environment. A catalog typically contains multiple schemas.

Column
A field or attribute in a database table. Each column in a table is given a name that describes the kind of data that exists in each row for that column.

COM
See Component Object Model.

Commit
To complete a database transaction, and write the changes to the database. *See also* Transaction.

Component Object Model
Microsoft's standardized model for creating component-based software. The COM architecture allows components made by different software vendors to be combined in a variety of applications. COM defines a standard for component interoperability that is not dependent on any particular programming language.

Cursor
A program's copy of the results of a database command or query. In ADO, a cursor is returned in the Recordset object. Cursors control record navigation, updatability of data, and the visibility of changes made to the database by other users.

Data Link
Contains the information needed to establish a connection to a data source. Data links are stored as text files with a UDL (Universal Data Link) extension. In AutoCAD, a data link is the same thing as a data source.

Data Provider
Software that owns data and makes it available to applications through OLE DB.

Data Source
A single database or database system. A single provider of data. *See also* Environment.

Data Source Name (DSN)
The logical name that allows a connection to an ODBC data source, such as an Access database. You set this name by using ODBC Administrator in the Control Panel.

Data Type

A constraint given to a single container of data that controls how the data is stored. Common data types in SQL are character, numeric, and date. VBA also uses data types, such as string, integer, and double.

Dock, Dockable

A term used in Windows to describe the behavior of a "child" window or toolbar. If a child window or toolbar is docked, it is attached to the frame of its parent window and moves and resizes with the parent window. If a child window is not docked, it is said to be "floating" and moves and resizes independently of the parent window.

Entity Handle

A persistent, unique name (hexadecimal number) given to every object that resides in an AutoCAD drawing file.

Entity Relationship Diagram

A database visualization technique used in the database design process that shows tables and their relationships to one another.

Environment

Also called a data source. The top-level component of a database system. The environment includes all the information needed to connect to a data source, including the server name, database name, username, and password.

ERD

See Entity Relationship Diagram.

Event

A mechanism used in COM libraries to notify an application when something occurs, such as the start or completion of a process, or a user's interaction with an interface component.

Field

See Column.

Foreign Key

A column used to establish a relationship with another table. The foreign key contains the value of the primary key in the related table. *See also* Primary Key.

Freestanding Label

See Label.

Index

A cross-reference to a particular column in a database table that provides the location of specific values in the table. When a query uses an indexed column, the database system can find the desired row much more quickly using the index rather than scanning the entire table.

Key

A column that uniquely identifies a row in a table. *See also* Primary Key.

Label

A multiline text object in AutoCAD that contains one or more field values from a linked table. A label can be attached or freestanding. An attached label is associated with another linked object and has a leader object that connects the label to the object. A freestanding label is an independently linked label object, and it is not associated or connected with any other object. In earlier versions of AutoCAD, a label was known as a *displayable attribute*.

Label Template

Defines the appearance of a label. The label template editor in AutoCAD is very similar to the multiline text editor. You can control all the properties of the text, such as height, font, and color. The label template also contains references to fields, which are then replaced with actual field values when the label is placed in the drawing.

Link

A connection made between an AutoCAD object and a row in an external table. The link is established when a copy is made of a key value in the table and that value is stored on the object.

Link Template

An object in AutoCAD that contains all the pertinent information needed to create links on objects to a database table. Each link template contains the data source, catalog, schema, table, and link columns used in the links. In earlier versions of AutoCAD, this was known as a Link Path Name (LPN).

Normalization

A database design technique in which tables are tested against a series of rules that help identify data redundancy and inefficiency.

Null Value

An empty field in a row of a table. Database systems distinguish null values from numeric values that are zero, or zero-length strings.

ObjectARX

AutoCAD's object-oriented development environment for C++.

Object Variable

A variable in a program that represents a reference to an object, rather than a simple data type such as an integer or a string.

ODBC

Open Database Connectivity. Microsoft's original database connectivity mechanism for Windows. ODBC provides applications with a common interface to communicate with a variety of database systems using the same set of tools, regardless of the nature of the system.

OLE DB

Microsoft's current universal data access standard. Developed to address some of the limitations of ODBC.

Primary Key

A column that uniquely identifies a row in a table. Primary keys are constrained by the database system to allow only unique values and to disallow null values.

Query

A request for information from a database. In SQL, a query is issued through the **SELECT** command.

Recordset

A cursor provided by ADO. The primary mechanism for navigating rows and editing data in an application. It is an object that contains a set of rows and the methods and properties with which to access and update the data.

Referential Integrity

A set of constraints established on a relationship between two tables that guarantees that the relationship is valid.

Relational Database

A method of database organization invented by E.F. Codd in which data are logically separated into tables. The tables can then be related to one another through the use of keys.

Row

A single unit of related information in a table. Also called a record. In relational terminology, a row is called a tuple.

Schema
An individual user's portion of a database. A schema contains a collection of tables for which access has been granted for the current user.

Shortcut Menu
A menu in a Windows application that is activated through a click of the right mouse button while the pointer is positioned over some object or area on the screen.

SQL
Structured Query Language. Pronounced "ess-cue-ell" or "sequel." A language used to formulate queries on relational databases.

Table
A collection of related information in the form of rows and columns. In relational terminology, a table is called a relation.

Transaction
A database term used to describe a modification or group of modifications performed on a database that can be reversed at any time, up to the time the modifications are committed to the database.

UDL File
Universal Data Link file. *See also* Data Link.

Visual Basic for Applications (VBA)
An API technology licensed by Microsoft to other software vendors. VBA is available in Microsoft Office products, as well as many Autodesk products such as AutoCAD and Autodesk World™.

Visual LISP
AutoCAD's LISP-like application development environment.

Wild Card
A character used to match one or more other characters in a query expression. The standard SQL characters are "_" (underscore) to match a single character and "%" (percent) to match zero or more characters.

APPENDIX

A

dbConnect Command Reference

DBCONNECT COMMANDS

Command	Menu Item	Description
dbcConfigure	Data Sources > Configure	Configures an external database for use with AutoCAD
dbcConnect	Data Sources > Connect	Establishes a connection to an external data source
dbcDefineLT	Templates > New Link Template	Creates a new link template in the current drawing
dbcDefineLLT	Templates > New Label Template	Creates a new label template in the current drawing
dbcEditLT	Templates > Edit Link Template	Edits an existing link template
dbcEditLLT	Templates > Edit Label Template	Edits an existing label template
dbcDeleteLT	Templates > Delete Link Template	Deletes a link template from the current drawing
dbcDeleteLLT	Templates > Delete Label Template	Deletes a label template from the current drawing

Command	Menu Item	Description
dbcImportTS	Templates > Import Template Set	Imports a set of templates to the current drawing
dbcExportTS	Templates > Export Template Set	Exports a set of templates from the current drawing
dbcPropsLT	Templates > Link Template Properties	Modifies the properties of a link template
dbcPropsLBLT	Templates > Label Template Properties	Modifies the properties of a label template
dbcExecuteQry	Queries > Execute Query	Executes a stored query
dbcNewQryTable	Queries > New Query on an External Table	Creates a new query in the current drawing based on a table in a database
dbcNewQryLT	Queries > New Query on a Link Template	Creates a new query in the current drawing based on the table associated with a link template
dbcEditQry	Queries > Edit Query	Edits an existing query
dbcDeleteQry	Queries > Delete Query	Deletes a query from the current drawing
dbcImportQS	Queries > Import Query Set	Imports a set of queries to the current drawing
dbcExportQS	Queries > Export Query Set	Exports a set of queries from the current drawing
dbcSelectLinks	Links > Link Select	Performs a Link Select operation

Command	Menu Item	Description
dbcDeleteLinks	Links > Delete Links	Deletes all links based on a particular link template from the current drawing
dbcExportLinks	Links > Export Links	Exports all links based on a particular link template from the current drawing
dbcLinkManager	Links > Link Manager	Edits the key values of a selected link
dbcReloadLabels	Labels > Reload Labels	Refreshes all labels based on a particular label template with new database table values
dbcShowLabels	Labels > Show Labels	Turns on label visibility of a selected label template
dbcHideLabels	Labels > Hide Labels	Turns off label visibility of a selected label template
dbcDeleteLabels	Labels > Delete Labels	Deletes all labels based on a particular label template from the current drawing
dbcViewTable	View Data > View External Table	Opens an external database table in View (read-only) mode
dbcEditTable	View Data > Edit External Table	Opens an external database table in Edit mode
dbcViewLinkedTable	View Data > View Linked Table	Opens an external database table in View (read-only) mode

Command	Menu Item	Description
dbcEditLinkedTable	View Data > Edit Linked Table	Opens an external database table in Edit mode
dbcExecuteQry	View Data > Execute Query	Executes a stored query
dbcSync	Synchronize	Detects broken links in the current drawing
dbcLinkConversion	Link Conversion	Converts links from previous releases to AutoCAD 2000 format

DATA VIEW COMMANDS

Command	Menu Item	Description
Command	Menu Item	Description
dvViewLObjects	View Linked Objects	Selects linked graphical objects when their corresponding records are selected
dvViewLRecords	View Linked Records	Selects linked records when their corresponding graphical objects are selected
dvAutoViewObjects	AutoView Linked Objects	Automatically selects linked graphical objects when their corresponding records are selected
dvAutoViewRecords	AutoView Linked Records	Automatically selects linked records when their corresponding graphical objects are selected
dvLink	Link!	Creates a link or a label
dvLinkToObject	Link and Label Settings > Create Links	Turns Link Creation mode on
dvLinkPlace	Link and Label Settings > Create Attached Labels	Turns Attached Label Creation mode on
dvPlace	Link and Label Settings > Create Freestanding Labels	Turns Freestanding Label Creation mode on
dvFind	Find	Searches for a value in the Data View window

Command	Menu Item	Description
dvReplace	Replace	Searches for and replaces a value in the Data View window
dvPrintPreview	Print Preview	Displays a preview image of a report in the Data View window
dvPrint	Print	Prints the contents of the Data View window to the current system printer
dvSettings	Options	Specifies Data View and query options
dvClearMarks	Clear Marks	Clears all marks from the Data View window
dvFormat	Format	Applies formatting to the Data View window display

APPENDIX B

Changes from ASE

INTRODUCTION

This appendix summarizes the changes in AutoCAD database connectivity features from previous releases of AutoCAD. The new dbConnect feature of AutoCAD 2000 completely replaces the older AutoCAD SQL Extension (ASE) found in Releases 12, 13 and 14.

The changes described in this appendix fall into three categories:

- Terminology Changes
- Command Changes
- Programming Interface Changes

TERMINOLOGY CHANGES

Several terms that are used to describe some of the features and components of DbConnect differ from those used in earlier versions of AutoCAD. Despite those terminology changes, the primary purpose of the feature remains the same.

ASE Term	dbConnect Term
Link Path Name (LPN)	Link Template
Displayable Attribute	Label
Key Column	Key Description or Key Value
Environment	Data Source

COMMAND CHANGES

All six of the ASE commands found in earlier versions of AutoCAD have been discontinued. However, all of the same functionality that was provided by those commands has been consolidated into the dbConnect user interface in AutoCAD 2000.

ASE Command	dbConnect Feature	dbConnect Menu Item
ASEADMIN	DbConnect Manager	
ASEEXPORT	Export Links	Links > Export Links
ASELINKS	Link Manager	Links > Link Manager
ASEROWS	Data View Window	View Data > Edit External Table
ASESELECT	Link Select	Links > Link Select
ASESQLED	Query Editor	Queries > New Query on an External Table

PROGRAMMING INTERFACE CHANGES

ASI AUTOLISP INTERFACE

ASILISP, the AutoLISP interface to AutoCAD SQL Interface (ASI), is still available in AutoCAD 2000. This interface provides AutoLISP programmers with the ability to communicate with databases from their programs. The ASILISP functions are defined in asilisp.arx, which must be loaded before any of the functions can be used.

In AutoCAD 2000, the recommended method of communicating with databases from an application is to use ActiveX Data Objects (ADO). ADO provides a great deal more power and flexibility than ASILISP. ADO is available to both VBA and Visual LISP, whereas ASILISP is only available to Visual LISP. It is for these reasons that discussion of the ASILISP function library is intentionally excluded from this book.

If you are looking for documentation on ASILISP, it can be found in the ObjectARX SDK documentation. The ObjectARX SDK can be downloaded from Autodesk's Web site.

AutoLISP programs that were written with ASILISP should still work in AutoCAD 2000 with a few exceptions, which are described below.

Discontinued Functions

The following ASILISP functions are no longer supported:

- asi_getcfg
- asi_setcfg
- asi_cmdtype

Changed Functions

The following ASILISP functions have changed slightly in AutoCAD 2000:

- asi_connect
- asi_alloc

The changes are described briefly below. For a detailed description of the functions, refer to the documentation.

asi_connect

The asi_connect function now takes the name of a UDL file as an argument, rather than an environment name. The new syntax for asi_connect is as follows:

(**asi_connect dataLinkFile** [*username*] [*password*])

The ***dataLinkFile*** argument is the name of the universal data link file that contains the connection information for a data source. This file is created by the dbConnect Manager when a data source is configured.

asi_alloc

The asi_alloc function now supports an optional updatability flag. The possible values of this parameter are "READ_ONLY", "UPDATABLE", or nil. The new syntax for asi_alloc is as follows:

(**asi_alloc stm_dsc cursor_name** [***scrollability***] [***updatability***] [***insensitivity***])

New Functions

The following ASILISP functions are new in AutoCAD 2000:

asi_databases

asi_infschema

asi_providers

These functions are described below:

asi_databases

(**asi_databases** ***providerName***)

Gets the cursor containing the databases that are available for the given OLE DB data provider. ***providerName*** is the name of one of the providers.

Shown below is an example of using asi_databases to get a list of available ODBC drivers. This function uses "MSDASQL" as the provider name, which is the OLE DB provider for ODBC data sources.

```
(defun C:ODBC( / ds ds_data )
  (cond
    ((null (setq ds (asi_databases "MSDASQL"))))
    ((null (asi_open ds)))
    (T
      (while (setq ds_data (asi_fetch ds))
        (princ (strcat "\n" (car ds_data)))
      )
      (asi_close ds)
    )
  )
  (princ)
)
```

asi_infschema

`(asi_infschema` ***envDesc rowsetName***`)`

Returns the cursor to one of the system tables of the Information Schema. ***envDesc*** is the SQL session descriptor. ***rowsetName*** is one of the predefined schema rowsets. In ADO, this is accomplished through the **OpenSchema** method of the **Connection** object. See Chapter 6 for more information on using **OpenSchema**.

asi_providers

`(asi_providers)`

Returns a cursor containing the system table of the OLE DB data providers. Shown below is an example of how asi_providers is used.

```
(defun C:PROVIDERS( / providers row_data )
  (cond
    ((null (setq providers (asi_providers))))
    ((null (asi_open providers)))
    (T
      (while (setq row_data (asi_fetch providers))
        (princ (strcat "\n" (car row_data)))
      )
      (asi_close providers)
    )
  )
  (princ)
)
```

ASI LINK AUTOLISP INTERFACE

The ASI Link AutoLISP interface is no longer available. This interface defined a series of functions (ase_xxxx) that could be used to access link information from AutoLISP. The Connectivity Automation Objects (CAO) component library replaces this functionality. CAO is described in detail in Chapter 7.

INDEX

Accumulate Record Set in Data View option, 34
Accumulate Selection Set in Drawing, 34
ActiveConnection, 175
ActiveX Data Objects, 2, 171-204, 216, 259
ActiveX Object Model, 172-73
AddNew, 180
ADE. (*See* AutoCAD Data Extension)
ADO. (*See* ActiveX Data Objects)
ADO library
  importing, 194-97
  referencing, 177-78
adBookmarkCurrent, 188
adBookmarkFirst, 188
adBookmarkLast, 188
adBSTR, 212
adChar, 212
adCmdStoredProc, 176
adCmdTable, 176
adCmdTableDirect, 176
adCmdText, 176
adCmdUnknown, 176
adEditAdd, 179-80
adEditDelete, 179
adEditInProgress, 179-80
adEditNone, 179
adLockBatchOptimistic, 176
adLockOptimistic, 176
adLockPessimistic, 176
adLockReadOnly, 176
adOpenDynamic, 175
adOpenForwardOnly, 175, 196
adOpenKeyset, 175
adOpenStatic, 175
adSearchBackward, 188
adSearchForward, 188
adVarChar, 212
adWChar, 212
Aggregate functions, 114-18, 259
Alias, 259
All, 123
Allow Docking button, 12
Annotation applications, 253-55
APIs. (*See* Application Programming Interfaces)
Application Programming Interfaces, 142, 171, 259
Applications
  developing, 168
  guidelines for a successful, 155-56
  identifying custom, 166-67
  off-the-shelf, 143
Apropos window, 195-97
As, 117-18, 121-22
ASE. (*See* AutoCAD SQL Extension)
asi.ini file, 63
Asset management, 150-51, 231-32
  designing, 156-68
Asynchronous execution, 174, 259
Attext, 145
Attributes, 1-2, 74-75, 79-80, 90-92
  block, 144
AutoCAD
  2000, 66-68
    objects, 172
  Data Extension, 146
  Database Connectivity Automation Objects, 172, 207-28
  SQL Extension, 1-3
    changes, 271-75
Automatically Pan Drawing option, 34-35

Automatically Store option, 45
Automatically Zoom Drawing option, 34-35
AutoPan and Zoom, 34-35, 51
AutoView Linked
  Objects option, 33
  Records option, 33, 35
Avg, 115, 118

Bill of Materials, 150-52
Block attributes, 144-45
Blocks, 1-2
Boolean operations, 48, 109-10, 259
Button, toolbar
  Allow Docking, 12
  Link, 14

CancelUpdate, 180
CAO. (*See* AutoCAD Database Connectivity Automation Objects)
CAO
  object model, 208
  utility functions, 241-43
Cascade option, 97-98
Case-insensitive string comparisons, 110-11
Catalog, 11, 54, 103-4, 260
  property, 211-12, 238
Cell Shortcut Menu, 26
Center option, 31. (*See also* Text alignment)
Changes from ASE, 271-75
Character data type, 105-6
Class example, 3
Clear property, 216
Close, 177
Column, 105, 260
  selection, 10
  Selector Shortcut Menu, 26
  shortcut menu, 10-11
  sorting, 10
  Values dialog box, 19, 40
  width, adjusting, 31
COM. (*See* Component Object Model)
Comma-delimited file, 50
Command, 192
Commands, command line
  dbcclose, 4
  dbconnect, 4
  DVLINK, 14, 17
  Export Links, 50
Commit, 28-29, 260
Component Object Model, 2, 171-72, 193-94, 260
Components, custom user interface, 142-43
Conditional expression, 108-11
Configure a Data Source dialog box, 7, 58, 103
Connection object, 174, 179, 192, 197
Connection, database, 6
ConnectionString, 174
conString, 195
Converting old links, 61-63
Copy, 36
Count, 114, 118
  property, 210, 212, 214, 216
Create, 106, 119
CreateLink method, 211, 219-21
CreateLinkExample function, 220
createLinkOnEntity function, 242
Criteria, 188
Cursor, 27, 260
CursorType, 175
Custom user interface components, 142-43
Cut, 36

Data
  Control Language, 106
  creation, 166-67
  Definition Language, 106-7, 119
  integrity, 147, 252-53
  link, 6, 260
  maintenance, 166-67, 250-52
  Manipulation Language, 106

modifying, 27-28
organization, 72-73
provider, 260
raw, 73
relational model, 74
retrieving, 176-77
source, 6, 104
available, 191-92
storage, 143-50
types, 105-6, 261
validation, 127-29
Data Link Properties dialog box, 7, 58, 66-67, 103, 234
Data Links subdirectory, 6, 8
Data Source, 11, 54, 260
Configuration utility, 3
Shortcut Menu, 67
Data Source Name, 260
Data Sources Node, 8
Data View
commands, 269-70
printing, 35-36
pull-down menu, 9-10, 24
toolbar, 24-25
window, 3, 9-11, 17, 23-36, 185
changing appearance, 30-32
Data View and Query Options dialog box, 33-35, 45
Database
connection, 6, 9
design, 71-88, 159-66
linking, 162-64
populating, 243-50
relational, 105
requirements, 76-77
storing drawing data in, 201-4
systems, 63
transaction, typical, 174-77
updating from drawing, 189-91
DataSource property, 211-12, 238
DataSourceLocation property, 209, 238
DataTypeEnum property, 212
Date data type, 105-6
dbcclose command, 4
dbConnect
command, 4
command reference, 265-68
components, 3
features, 3
invoking, 3-4
Manager, 3-5, 7, 181
object, 208-12, 214, 216
property, 210-12, 214-15
DBQ files, 17, 45
DCL. (*See* Data Manipulation Language)
DDL. (*See* Data Definition Language)
Define Name dialog box in Excel, 65
DefinedSize property, 212, 238
Delete, 101, 106, 126-28, 192
method, 222-24
DeleteLinkExample function, 224
Delete Link Template, 56
Design validation, 98-99
Dialog box
Column Values, 19, 40
Configure a Data Source, 7, 58, 103
Data Link Properties, 7, 58, 66-67, 103, 234
Data View and Query Options, 33-35, 45
Format, 32
Import Query Set, 45, 103
Label Template, 15-16
Layer Properties Manager, 232-33
Link Conversion, 62
Link Select, 47-48, 50
Link Template Properties, 54-55
Link Template, 12-13, 54-55, 58, 234, 237
Macros, 183
New Label Template, 15
New Link Template, 12-13, 234-35
New Query, 18, 37-38

Options, 68-69
Query Editor, 19, 38-43
Range Query, 21
References, 177-78
Replace, 28
Sort, 10-11, 30
Synchronize, 59
Warning Message, 44
DML. (*See* Data Control Language)
Dock, 261
Dockable, 261
Document property, 211-12, 214-15, 224
Drag and Drop, 46, 57
Drawing
upgrading from the database, 186-88
Drop command, 106
DVLINK command, 14, 17

Edit
buffer, 179
mode, 27
Edit Link Template, 53, 55
EditMode property, 179-80
Editor, Query, 3, 17-21
dialog box, 19
EED. (*See* Extended entity data)
Entities, 74-75, 77-78, 89
Entity
handle, 261
relationship diagram, 74-88, 102, 261
Environment, 103-4, 261
EOF, 178
ERD. (*See* Entity relationship diagram)
Err object, 225
Error
object, 216
trapping, 197-98
ErrorCode property, 216
ErrorDescription property, 216
Errors collection, 208, 215-16, 225-26
Event, 261
Examples, real world, 2-3
Execute, 192-93, 249
Export Query Set, 45
Export Template Set, 57
Exporting links, 50-53
Expression, conditional, 108-11
Extended entity data, 144-46
External storage, 143-44, 147-49

FieldName property, 212, 238
Fields, 105, 261. (*See also* Attributes)
File-level locking, 148
Find, 188
Font characteristics, changing, 32
For Each
construct, 210
loop, 191-92
Foreign key, 80-83, 92-94, 261
Forms, normal, 83-87
Freeze option, 11, 31
From, 113-14, 118, 120, 132-33
Functions, aggregate, 114-18

Get function, 196
getAdoConnection function, 242
getCenterOf function, 243
getConnectionString function, 242
getDbConnect function, 242
GetErrors method, 209
GetLinkTemplates
function, 242
method, 209
getLinkValueFromEntity function, 242
GetLinks method, 209, 214, 216-19
getPolylineFromPoint function, 243
getSpacePolylineSelection function, 242
getTextInsidePolygon function, 243
Grant, 106
Graphical
Selection, 47
user interface, 3

Group, repeating, 83-84
Group By, 115-16
GUI. (*See* Graphical user interface)

Hatching spaces, 255-58
Having, 116-17
Hide option, 11, 31

Icon
  dbConnect, 7
  ellipsis, 8, 181
  Return to Query, 43
  View Linked Objects in Drawing, 32
  View Linked Records in Data View, 33
Import Query Set dialog box, 45, 103
Import Template Set, 57
In, 124
Index, 262
Information, 73
Initialize event, 191-92
Insert, 101, 106, 124-26, 192
Integrity
  data, 147
  referential, 96-98
Interface, creating the, 138
Internal storage, 143-46
Interview process, 76-77, 159
IsInside function, 243
Item method, 210, 212, 214, 216
Iterative Querying, 46-47

Join tables, 112-14
Justification, text, 31

kAttachedLabelType, 217
kEntityLinkType, 217
Key, 11, 262
  column(s), 11, 54-55
  foreign, 80-83
  identifying, 92-94
  primary, 80-83
  value(s), 11
KeyDescriptions
  collection, 212, 219
  object, 212-14
  property, 211-12
Keys collection, 219
KeyValues
  collection, 221
  object, 226-28
  property, 215, 219
kFSLabelType, 217

Label
  button icon, 16
  Template dialog box, 15-16
Labels, 133-37, 262
  attached, 14
  creating, 14-17
  freestanding, 14-17
  shortcut menu, 59-60
  template, 15, 262
Layer Properties Manager dialog box, 232-33
Left option, 31. (*See also* Text alignment)
Like operator, 110-11
Link, 11, 262
  button, 14
  converting old, 61-63
  exporting, 50-53
  object, 215
  Path Name, 61
  Select
    dialog box, 47-48, 50
    feature, 46-50
  shortcut menu, 59-60
  synchronizing, 58-59
  template, 12, 262
    deleting, 56
    dialog box, 12-13, 54-55, 58, 234
    modifying, 53-56
  to AutoCAD, 129-31

Link Conversion dialog box, 62
Link Template
  collection, 208-11, 217
  dialog box, 12-13, 54-55, 58, 234, 237
  Properties, 53, 58
Linking
  objects, 11-14
  polylines, 243-48
  schemes, 152-55
Links collection, 208, 214-15
LinkTemplate
  object, 211-12, 216-17, 219, 225, 239
  property, 215
LinkType property, 215
LinkTypes argument, 217
List Box object, 191-92
Locking, 148
LockType, 176
LPN. (*See* Link Path Name)

Macros dialog box, 183
Maintenance effort, 73
Mark Indicated Records option, 34
Marking Color option, 34
Max, 115
MDAC. (*See* Microsoft Data Access Components)
Metadata applications, 152
Microsoft Access, 27
Microsoft Data Access Components, 8
Microsoft Excel, 36, 63-65
Min, 115, 123
MoveFirst, 176
MoveLast, 176
MoveNext, 176, 178, 196
MovePrevious, 176
MTEXT, 15
Multiline text object, 14
Multiple Link Templates, 47
Multi-User support, 148

Name property, 211-12
Needs assessment, 157-59
Nested query, 123
New Label Template dialog box, 15, 234
New Link Template
  dialog box, 12-13, 234-35
  option, 12, 131, 136
New Query dialog box, 18, 20, 37-38
No restrictions option, 97
Non-graphical information, 144
Normal forms, 83-87
Normalization, 83-88, 94-96, 164-66, 262
Nouns, 77-78
Null Value, 262
Numeric data type, 105-6
NumericScale property, 212, 238

Object
  data, 144, 146
  variable, 263
Object shortcut menus, 59-60
ObjectARX, 171, 263
ObjectId
  parameter, 219, 224
  property, 215
ObjectIDs array, 217
Objects
  attached label, 14
  freestanding label, 14
  linked, viewing, 32-33
  linking, 11-14
  Multiline, 14
ODBC. (*See* Open Database Connectivity)
  Data Source Administrator, 65
  Microsoft Excel Setup dialog box, 66
Off-the-shelf applications, 143
Office example, 3
OLE DB, 2, 263
Open method, 174, 198
Open Database Connectivity, 2, 6, 64, 263
  data source creation, 65-66

Open Tables in Read-Only Mode option, 69
openConnection function, 242
openRecordset function, 242
OpenSchema, 192
Options, 174, 176, 192
  dialog box, 68-69
Oracle, 63, 68
Order by, 112

Parcels example, 2, 6-9, 11-14
Password, 174
Paste, 36
Precision property, 212, 238
Preferences, 191-92
Primary key, 11, 80-83, 92-94, 105, 263
Print
  Data View icon, 36
  Preview, 36
Problem statement, 77, 157
Process, interview, 76-77
Put function, 196

Query, 107, 263
  applications, 253-55
  Builder, 17, 36-43
  Editor, 3, 17-21
    dialog box, 19, 38-43
  executing a, 175-76
  nested, 123
  processing options, 45
  sharing, 45-46
  SQL, 101
Quick Query, 17-20, 36

Range Query, 17, 20-21, 36
  dialog box, 21
Real world examples, 2-3
Record, 105, 146
Record Indication Settings, 34-35
Record-level locking, 148
Records, linked, viewing, 32-33
RecordsAffected, 192
Recordset, 175-80, 188, 238-39, 263
  retrieving, 199-201
References dialog box, 177-78
Referential integrity, 96-98, 263
Relational
  data model, 74
  database, 263
Relations, 74
Relationships, 74-75, 78-79, 89-90
ReloadLabels method, 209, 224-25
ReloadLabelsExample, 225-26
Replace dialog box, 28
Reporting, 166-67
Reports, producing, 73
Requirements, user, 88-89
Restore option, 28
Restrict option, 97
Result set, 48
Revoke command, 106
Right option, 31. (*See also* Text alignment)
Row Selector Shortcut Menu, 26
Rows, 105, 263
  appending, 28-29
  deleting, 29

Schema, 11, 54, 103-5, 192, 264
  property, 211-12, 238
SearchDirection, 188
Security, 149
Select Indicated Records option, 34
Select, 101, 106-8, 118, 120, 124-26, 175
SelectObjectsByKey function, 228
selectPolylinesFromPoint function, 242
SelectPolylinesOnScreen function, 242
Send as Native SQL option, 45
Set, result, 48
SetEnclosed function, 250
SetHasWindow function, 250
Shortcut menus, 25-26, 264

Show All Records option, 34
Show Only Indicated Records option, 34-35
SkipRows, 188
Some, 123-24
Sort dialog box, 10-11, 30
Sort option, 10-11
Sorting output, 112
Source, 175
Space-delimited file, 50
SQL Query, 17, 36, 43, 47
SQL. (*See* Structured Query Language)
Standard option, 31. (*See also* Text alignment)
Start, 188
Statement, problem, 77
Storage of data, 143-50
Store Links Index in Drawing File option, 68-69
Store, 47
String comparisons, 110-11
Structured Query Language, 2, 101-39, 264
Subquery, 123
Subset, 108
Sum, 115
Supports, 179
Synchronize, 58
 dialog box, 59

Table, 11, 54, 74, 83-84, 103-5, 264
 getting a list of, 192
 joining, 112-14
 linking, 93
 primary, establishing, 160-61
 property, 211-12, 238
 querying from multiple, 112-14, 132-33
 Selector Shortcut Menu, 25
 virtual, 118
Templates Menu, 53-54
Text alignment, 31. (*See also* Justification, text)
Tool bar, 4
 Data View, 24-25, 32-33
 Shortcut Menu, 25
Tools menu, 4-5
Transaction, 264
 model, 147-48
Tree view, 4
Type property, 212, 238

UDL files, 6, 8, 174-75, 191, 264
Unhide All option, 11
Unordered, 105
Updatable property, 215
Update, 101, 106, 126, 180, 192, 249
 method, 221-22
UpdateLinkExample function, 222
UserID, 174
Utility, Data Source Configuration, 3

Validation, design, 98-99
VBA. (*See* Visual Basic for Applications)
Version property, 209
View, 118
 mode, 27
 Table option, 12
Virtual table, 118
Visual Basic for Applications, 2, 171-72, 209, 264
Visual LISP, 2, 171-72, 193-204, 264

Warning Message dialog box, 44
Weak entities, 75
Where, 108, 113-14, 117, 121-22, 126, 188
While/Wend construct, 178-79
Wild cards, 110-11, 264
Window Shortcut Menu, 25
Window, Data View, 3, 9-11, 17, 23-36
 changing appearance, 30-32

Zoom Factor option, 34
Zoom, AutoPan and, 34-35, 51